Control

The Foundation of Life

Exploring the significance of a simple concept to aid in understanding what we perceive about us.

Control

The Foundation of Life

Lance Packer

To my wife, Ginny, who revealed the pervasive requisite
of memory in all that makes us human
and gave me a lifetime of togetherness to always
cherish.

CONTENTS

INTRODUCTION

What is the motivation for action? That is, action by oneself, other individuals, societies, and social groups, even actions by other mammals, any animal, even plants and bacteria—by any living thing? What is it that makes us do what we do? It would seem that knowing the answer to this question would provide valuable knowledge. In fact, such a quest would appear to lie at the basis of all understanding of human history and explanation of current human activity, and the basis for understanding all life forms since the goal of science is explaining what makes the world operate. But then, isn't this all rather obvious? Yes, and the obvious is often overlooked.

Perhaps using personal experience as an example would be useful. In the later years of my formal education in anthropology, after having studied many cultures of the world and tried to more clearly understand what it meant to personally share any culture, the issue gradually focused more on why there were so many different cultures rather than on the particular characteristics of each. Those specific characteristics were fascinating enough, but it begged the question of why there were different ways to, for example, formalize a marriage, recognize adulthood, cook a meal, clothe oneself, or communicate with each other. What did each human culture share? What did all humans share? Was it some universal need, such as food, sexual regulation, and protection from the elements? Certainly much has been written about this and lists have been made. Yet, there's always something that fits on one list and not another or needs to be added, subtracted, or merged.

Such precariousness in list-making did not seem to fit my sense of a universal explanation for why cultures are different, or why they exist at all. It would appear that there isn't anything genetic that specifically dictates that a human culture needs to exist. But why stop there? Isn't the question much broader than asking what is the motivation for behavior in just human culture? Shouldn't it be asked more universally for all life forms—and that human culture is subsumed as part of a broader concern To me, it was more conceivable that an answer should apply to all life, not just humans. Now, that would be something worth investigating further.

After leaving my time of formal education and spending many years teaching in public schools, while not actively pursuing this question of a universal explanation for motivation of action, it nevertheless was always in the back of my mind and occasionally contemplated as life's events came to touch on its relevance for explaining what I observed in daily life. In particular was the behavior of students in the classroom.

One of the primary goals of a classroom teacher is to maintain control of student behavior so that another primary goal, learning by the students, can occur. Like other teachers, I had my system set up for achieving both goals, which was rationally structured to allow specific levels of student choice matched with required activities. Generally, it was successful. However, some students, because of cognitive inability, emotional upset, family problems, and so on—whatever it was for that day—had trouble with my system. In other words, they "got in trouble": talking excessively, making noises, refusing to attempt the work, and being physical with other pupils, to name a few. Constantly, I was confronted with the necessity of keeping students on task and solving problems caused by those who got into trouble.

After a few years of this daily teaching demand, I started to ask myself: Seriously, why do kids get into trouble? It got a bit tiresome to deal with, especially since the logic of my teaching structure was supposed to guarantee students that they would learn with as little conflict as possible. What was at work here? What was the motivation for action?

A common response, of course, is that trouble-makers are just ornery kids, or mean, stupid, bad seed, evil, or whatever popular reason can be thought of. I rejected all these because they offered no opportunity for remedy; they were all reasons for immutability. But realistically, why try? Just kick the kid out. Yet, that was a solution I could not accept because the trouble-makers weren't always a problem; they did well some days, perhaps not as well as others, but clearly, there was some evidence of ability. How to get to them was the problem. And it seemed that accessing the positive side of the child meant understanding what motivated them, what caused their seemingly unreasonable behavior.

Another explanation, rather than inherent inability and evilness, could be that their family had problems that sparked the trouble: financial instability, lack of supervision, anti-social modeling, dysfunctional relationships, and many other options. Or it could be more immediately personal: a friendship turned bad, an argument with another teacher, a stomach-ache, the flu, a disappointing test score, and a dozen more possible events of the day. Or it could be cultural: an immigrant struggling with the language, religious beliefs, values toward education or certain subjects, pressure from parental expectation, and on and on.

So at the end of the day, how was I, as a teacher, to assess what had happened during that time and what my responses had been—and then get up the next day and try all over again, without beginning to resent the trouble-makers even more?

One realization I came to early in my teaching career was that with rare exception, the acting out of the student rarely had anything to do with me personally: I was just the adult in the way of responding to whatever was bothering them and therefore faced the substance of their action. This would usually become apparent after we had a chance to talk privately about the incident, or simply after a short passage of time. That understanding helped me direct my immediate response to the situation and allow the class to continue with as little disruption as possible. Of course, for future reference at some point, it was necessary to further investigate the parts of the conflict that occurred, which could be rather complex, as suggested.

Another and more important, realization I came to because of these conflict situations was that every student in my classes wanted to be successful, regardless of how it seemed on the surface. Given a new task, they would all try to do what I required, which was obvious from their effort, expressions on their faces, and their comments. Great! That's what I hoped for.

However, in every situation, some students would soon have difficulty accomplishing what I expected and get into trouble doing something they shouldn't. It wouldn't be the same kids every time but would vary with the assignments, clarifying that if I varied the exercises enough, all students should have a chance to be successful. So, variation in assignments to meet the different learning styles of all students was what they needed. Nevertheless, as much as I tried every day, I couldn't provide the variety needed for each assignment so every student could always be successful, at least successful in doing what I expected. Some still were motivated to take action I deemed disruptive and unsuccessful.

Then it finally dawned on me. In the disruptive student's mind, he was being successful—just not according to my definition. If it was getting laughs from the other students, he was successful. If it was avoiding having to do something he wasn't capable of (for whatever reason), he was successful. If it was displacing anger with his parents onto an adult, such as me, who was also requesting compliance, he was successful. In other words, every student wants to be successful and will find one means or another to achieve success, whether in an educationally acceptable form or another not so acceptable form.

This realization put student behavior on a new level of understanding which not only aided my personal relationship with students but also helped educationally in trying to establish a learning environment that took into account the individual students' preferred learning styles, and the demands of their personal life that day. None of us has a great day every day, and students have more trouble with that as they grow up. However, just being aware of the issue, and making acceptable accommodations, goes a long way toward accomplishing the primary goal, in this case, of providing an optimal learning environment in the classroom.

Great. But the question still lurked in my mind: Is there some universal factor behind all this classroom conflict and struggle to be successful, something that can more broadly explain not only student behavior but also human behavior in general? After all, don't all humans want to be successful in their actions, legal and criminal? Why else would people try so hard, for example, to raise a family or increase their standing in their profession or, for another kind of example, to develop a narcotics network or increase the respect they have in their gang community? What is it that motivates people to strive for success, whatever the definition? Since the definition of success varies from society to society, and even within each society, it would seem that there must be some kind of universal motivation common to all humanity that accounts for this behavior. But what is it?

Eventually, I determined that perhaps success wasn't quite the concept that best described what I observed since it has subjective value connotations. I needed a broader concept, and the descriptor "control" seemed to best fit what I was searching for. What my students were seeking was as much control of their situation as they could get, given the bounds of culture and society. Those who went along with the control system for learning that I had set up received the customary reward of grade and teacher/student approval. Those who didn't go along with it didn't receive approval and got in trouble.

Yet, the troublesome student wasn't necessarily unsuccessful in his attempt to assert control over his situation; it just wasn't what I wanted as a result. The issue was one of control, between the students and me, and there were several possible outcomes: some socially acceptable according to adult culture, some socially acceptable according to student culture, and some not acceptable to either.

Putting the teaching situation in the context of a struggle for control and acceptance or rejection of that control made it easier for me to analyze student interactions not only with myself but also with each other. Besides, it removed the tendency to explain student behavior as due to innate, unchangeable character; instead, behavior could be explained by analyzing the attempts to assert control in a particular interaction.

So, for example, when a student stops trying to work on an assignment and starts poking the kid in front with his pencil, the teacher should try to figure out what the student is trying to control. First, he is rejecting the existing control of the teacher and learning structure. The question would be how the structure could be changed so the student would feel he has greater direction over it (changing the level of difficulty, partnering with another pupil, etc.). Second, the trouble-maker's attempt to assert some physical influence over the kid in front must be dealt with immediately. This is where the teacher's box of tricks to maintain discipline comes in; it's strictly a matter of asserting teacher control to stop the student trying to assert his control so the disruption can be immediately stopped and then turning to redirect the power the student is seeking.

Ultimately, having the student be able to assert some control over his learning parameters of the assignment ("buying into it") because of adaptations made by the teacher is the best means of removing the unacceptable behavior, and more lasting. The alternative is outright physical and/or verbal struggle for dominance, with neither student nor teacher seeking adaptations and alternatives, resulting in punishment—and resentment by both sides—with any return to an optimum learning environment jeopardized.

In essence, was control the concept I was seeking which best explained the motivation for action by humans? It seemed to make sense for my teaching situation, and it seemed to fit for other human behavior that I could see about me in my personal life of family, teaching staff, and community interaction (such as simply paying for items at a store and obeying stop signs), and elsewhere. If control was the key in my small part of the world, then it would be the same for all human interaction, a universal explanation for all behavior.

But maybe it's more than that. Since humans are part of the animal world, could it also be true for all animals? And for plants too? Perhaps a universal explanation for the motivation for the behavior of all living things? I believe that it is, and the rest of this treatise examines that suggestion concerning all life forms, and particularly human, to see if control is the broad conceptual paradigm that it appears to be, encompassing all the processes and elements of life.

In conclusion, my understanding of the concept is based on personal life experience, as depicted previously, but also on examples of world history, science, and every other existing explanatory tool of human knowledge that I have delved into throughout my life. Therefore, I consider the reader to be just as capable as the author in finding relevant examples needed to help explain the universality of the paradigm beyond those I have provided in this text. The entire history of and current knowledge about the world, and one's personal experience, can provide a reader with all the instances useful for further insight.

Since the focus of this endeavor is to outline the paradigm and to enhance one's understanding of it, I further encourage the reader to also use every opportunity available to apply the concept to personal life situations and assess its relevance. I believe it will then become clear that there are many more associated areas of human behavior to be considered than what I have touched upon. As with any useful philosophical concept, the test of its value is in its application to the life of the individual and those social groups to which he or she belongs. I hope this effort meets that pragmatic standard.

I

BASIC PRINCIPLES OF CONTROL PARADIGM

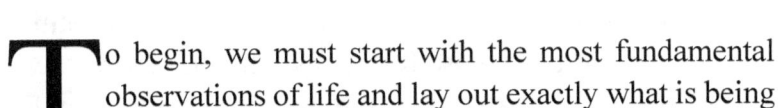

To begin, we must start with the most fundamental observations of life and lay out exactly what is being controlled: that is, the essential elements of the environment in which life occurs and how they interact.

Attributes of the Environment

The environment of all living beings consists of:
1. non-living chemical elements and physical planetary forces that act upon them (geologic, atmospheric, hydrologic, weather, astronomic, gravity, etc.) and
2. other living beings (any life form that is capable of cellular function, growth, and reproduction).

Furthermore:
1. Everything in an environment is interrelated and contingent, and all environmental necessities for life are therefore contingent on each other.

2. Control of the environment is necessary for a life form to exist.

3. To continue existing, a life-form must be able to perceive the environment and establish boundaries that define the extent of as much environmental control as possible.

4. However, life forms are unable to control every environmental factor due to the quantity and attributes of environmental factors and competition for control of those factors by other life forms.

5. Since all parts of a life form's environment are interrelated and constantly responding to every other part, change and adaptation of control boundaries are continually necessary for it to survive.

6. This ability of a life form to adapt its environmental control boundaries is made possible by a combination of genetic and learned information. Over time, the evolution of genetic content and capacity for learning (and also, mainly for humans, the transmission of learned information via culture) allows descendants of individuals to successfully establish their needed boundaries. Thus, a species can adapt to a particular, but changing, environment.

7. In addition, an individual life form is the primary and essential element of the living component of the environment since it's the component that seeks control of other environmental elements. Besides,

any group action by individuals can affect an individual's control but is secondary in importance. Simply stated, objects which can't control their environment are what we call inanimate, in contrast to animate objects. It's the crucial distinction that makes one clump of chemical elements into a rock or a lake and another clump of elements into a tree, a worm, or a human being. A living entity consists of a specific organization of chemical elements into a form that grows, reproduces, and tries to control as much of its environment as needed and possible. Once a threshold level of control is lost, it ceases to live— it's dead—and the once-living organization of elements again reverts to its previous elementary status and becomes inanimate again. Thus, death is defined as an animate being losing essential minimal control of its environment.

In contrast, inanimate objects such as a rock, can't actively control their environment, living or inanimate. They can only be acted upon by living beings or natural forces of their environment. For example, a human can use the force of gravity to roll a log down a hill or use wood to build a house to take shelter from cold temperatures. It can also take a rock and chip it until it becomes a spear point or crush it in a mill to separate out the iron for making a refrigerator. In a similar manner, various species of birds and mammals also eat certain types of clay or rock to extract usable minerals for their diet. In contrast, gravity will cause a rock to roll down a steep slope if it's unstable and freezing/thawing cycles will cause water to crack the rock into smaller pieces. The rock itself, however, can't do anything on its own to control gravity or freezing; nor can it prevent the humans or plants in its environment from using its constituent properties for their purposes. By the same token, a force, such as gravity, can't redirect its attribute of attraction and send that rock rolling back up a steep slope.

Since control is a requirement of living beings in order to be alive, this means not only conflict with inanimate objects and forces but also between individual beings and groups of beings. This is true for humans as well as all forms of life, whether animal, plant, or bacteria (to classify them simply). All have to control their environment to survive and that means inevitable conflict. How that conflict occurs and is handled by each species' control processes is really the unique essence of the species, and it's what scientists study so keenly and each species also carefully learns or genetically adapts to through evolution to improve its survival chances. The crucial point is that conflict is inevitable, and the history of that conflict activity is recorded as genetic evolutionary changes and/or by an individual's cognitive processes which are imprinted as memory in a brain (as in a hungry lion), and also for humans as a cultural artifact (such as clay tablets, paper, or digital storage).

Viewed as a process in time, life could be said to be a transitional event going from the inanimate to animate and back again. All living organisms share this pattern, and many branches of science have been developed over the centuries to detail these processes of transition. Medical science, in particular, is concerned with the inception of life, the systemic forces that continue life, and the failure of systems leading to death. The goal of medical research and practice is to help individuals, especially human, maintain control of the multitude of factors which sustain their lives. Ecological sciences, in essence, have the same goal except that the focus is on discovering the control boundaries and interactions of the total environment for all life forms and seeking to provide information that can influence the balance of these according to social values.

Interaction of Non-living and Living Environments

Control boundaries are also influenced by other environmental elements besides genes, learning, and culture. Non-living elements such as water, atmosphere, landmass composition, and distribution as well as the forces of nature such as weather, climate, volcanic eruptions, continental drift, gravity, and chemical reactions interact not only with each other but also with living beings. Before the appearance of life on this planet, these forces and elements were the sole determinants of change on Earth. For example, planetary magma movement causing continental formation and drift led to volcanic eruptions which massively changed the geologic and geographical features on the Earth's surface. Water and wind later helped sculpt those features over the ensuing eons.

Then after the appearance of life, living beings were affected by those same natural forces and elements, and because of the inherent nature of their need for control, they increasingly affected that natural environment directly and indirectly from the interaction of life forms with each other. For example, primitive cyanobacteria were early direct contributors to sedimentary stromatolite rock formations, as were coral polyps which added limestone reefs of vast quantity to the landscape. Indirect contributions came from the control competition between plants and animals, as in overgrazing by wildlife and domesticated livestock and over-cultivation leading to massive erosion of soil and desertification. Another, more pervasive, example is the effect that human activity has had on global climate change with increased production of greenhouse effect gases, resource utilization, and population growth. Geologic, historical, and ecological studies of life prior to and throughout human existence provide a multitude of examples demonstrating this competition for control by life forms of the living environment and their effect upon the non-living environment.

The essence of all these confirms that there's no such thing as a non-control aspect to the environments, non-living or living. All are tied up in constant change: Planets don't exist except for the constant interaction of the basic forces and elements of the universe, and life doesn't exist except for the constant control-seeking that is essential to meet basic requirements. To paraphrase an axiom by Heraclitis, there is no constant except constant change; and we must take it that this is the starting point for understanding life and the world in which it exists.

Unfortunately, that simple fact is very frequently forgotten in human social policy discourse for making decisions about these all-important environmental elements; the current status is often assumed to last indefinitely, and plans are made on that basis. Change and control must be the framework for thinking about how to structure any such discussion and what the results of action might be; not doing so ends in temporary solutions and disappointment when inevitable unexpected changes occur.

The Primacy of the Individual

A final important principle of the control paradigm is that the individual living being, in contrast to any group to which it belongs, is the most important part. The essence of living is the individual. A group doesn't live; an individual does. Furthermore, the process of grouping is one of perception. This includes not only a social group, in whichever species it may exist, but also a genetic grouping; neither exists except in the mind of those doing the grouping. In a genetic grouping, the genetic coding of the individual's cells allows that individual to live, to enable reproduction of the species, and to provide genetic variation which may affect the individuals' ability to adapt to changes. A social grouping relies upon the interactive behavior of individuals to survive through mutual defense, food-finding, sexual reproduction, etc. In both, it's the individual that makes the perceived grouping possible: no individual, no group. This is the case for all life forms, simple or complex; and with the level of complexity of human physiology and behavior, the primacy of the individual, in that case, is even more complicated.

Because culture is an exceedingly powerful control process, when a group of animals shares a culture, such as humans do, the strength of social bonds is enhanced by culture to a high degree and often seems to surpass the value of the individuals in the group. This can be manifest particularly in warfare, where ideas of sacrifice for the tribe or nation-state surpass all other considerations. "Cannon fodder" is an old but useful shorthand image of what this means for the individual, and seems timeless in human history through the present. Similar to this but not as drastic are ideas based on saving an ethnic heritage, a religious belief, a language, or a culture itself. In all these, the social group is repeatedly considered to be more valuable than each group member.

What is often ignored is that all these human grouping ideas are based on the concept of culture and are thought of as if they exist somehow by themselves; whereas in fact, they don't exist at all except in the minds of those who share them as part of a culture. They are all invisible cognitive constructs exhibited only through behavior and products of that behavior. Remove the behavior and they cease to exist; remove the individuals and they cease to exist.

Moreover, a culture and its artifacts continue to exist only if individuals find them useful. Our museums and antique stores are full of old cultural artifacts, interesting but useless in practicality. For the same reason and as an example, languages have always disappeared throughout history because they no longer meet the needs of the individuals who once used them, despite the desire of the remaining elderly speakers who view them as key to maintaining an ethnic identity for younger generations.

Viewed from a more encompassing perspective, cultures and their elements disappear because the control parameters of the society's individuals have changed. If social relationships between human groups change due to excessive warfare, for example, the members of both groups may find it advantageous to intermarry and thereby increase their bonds of cooperation, build up their population to stave off threats from other social groups, and consequently increase the influence of the newly expanded tribal or national territory.

Or maybe that volcano that just erupted puts old rivalries into a new perspective, and basic survival by aiding each other becomes a more immediate interest. Archaeology and history provide a never-ending story of control boundary adaptations. It can be imagined that those other living beings such as cetaceans and primates who have at least some cultural ability may likewise have a history of such adaptations, but it's likely limited to memory-based behavior and oral transmission of some type.

The point of all this is the primacy of the individual, and not the group, because each individual alone carries the species' genetic code, can learn, and, especially with humans, also transmit culture. It's the individual that ultimately establishes control boundaries to meet its needs.

Additionally, it must be remembered that an individual's control depends on how it responds and adapts to the total environment, not just a specific part—which includes other living beings, bacteria, plant, and animal, as well as all aspects of the non-living world such as weather, rocks, minerals, crashing meteors, plate tectonics, temperature, tornadoes, and more.

The essence of the composition and interaction of all these are the stuff of science, in addition to religion, mythology, philosophy, and other human systematic efforts of explanation—and control—of what we perceive about us in this world.

As a way of illustrating the primary structure of the control paradigm, a graphic outline may be useful:

Outline of Control Paradigm

FUNDAMENTAL PRINCIPLE OF CONTROL

All life forms must control as much of
their environment as possible to survive,
and this requires change and adaptation.

PRIMARY PROCESSES OF CONTROL

Evolution:

The adaptive changes to life forms over time
due to conflicting results of efforts to control
elements of the non-living and living environments.

| --Structural-- |
| Genetic DNA and RNA transmission over generations through reproduction; physiological development through adaptation; increased neurological complexity. |
| --Behavioral-- |
| Utilization of neurological complexity and brain functioning for consciousness, sensory perception and interpretation, memory, and decision-making; utilization of diverse physiology and activity. |

Ecology:

The interactive relationship between individual life forms
and their non-living and living environments.

--Non-living Environment--
Natural chemical elements and planetary forces that act upon them: geologic, atmospheric, hydrologic, astronomic, weather, gravity, etc.
Living beings interact with them to utilize the materials and forces available and to protect themselves from adverse results of interaction: e.g., rocks and soil for construction and protection from landslides, floods, weather, wildfire, etc.

--Living Environment--
Any life form that is capable of cellular function, growth, and reproduction: all species of bacteria, plants, and animals, including humans.
Living beings interact both within and between species through conflict or cooperation with other life forms for food sources, defense of territory, access to materials to utilize, reproduction of the species, etc.

Culture:

The development and transmission of learned behavior and concepts from one individual and generation to another. Largely used by humans but not exclusively.

--Behavior--
Ritual, language, food procurement, play, sexual activity, entertainment, instruction, expression of emotion, fighting, utilization of natural resources, artifact production, etc.

--Concepts--
Rules and laws, government, kinship, family, voluntary
social groups, education, artistic standards and style,
religion, beliefs, mores, philosophies, economics, etc.

All three processes interact with each other contingently, one affecting the other in response to actions taken by an individual life form to control its environments. Over time, adaptations from control attempts result in structural and behavioral evolution to improve the life form's success in control of the environments. Ecological relationships also change and require further adaptions for control. To meet the need for change, neurological complexity increases dramatically through evolution, resulting in even greater control potential for some species. With the human species, in particular, physiological and neurological evolution has generated the capacity for culture, which allows accelerated and far-reaching control potential and actualization.

FURTHER EXPLANATION AND EXAMPLES

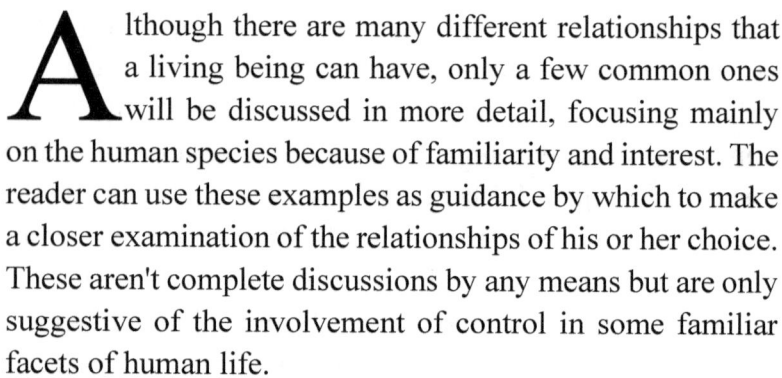

A lthough there are many different relationships that a living being can have, only a few common ones will be discussed in more detail, focusing mainly on the human species because of familiarity and interest. The reader can use these examples as guidance by which to make a closer examination of the relationships of his or her choice. These aren't complete discussions by any means but are only suggestive of the involvement of control in some familiar facets of human life.

The Individual

To use the human example, each person continually struggles throughout their life to assert control over the environment to meet his or her drives, be they basic needs such as food and protection or desires such as material wealth and fame.

Determining exactly what is a need or a desire has been the subject of thoughtful discourse over thousands of years of human thought and writing and isn't to be resolved here. Individuals attempt to control as many aspects of their environment as they can because it's at the core of what it means to be alive.

Beginning with our conception, we each seek to receive the nutrients we require for cellular growth and maintenance; and after birth, we add the interpersonal relationships which keep our physical condition healthy as well as that of our psychological and social circumstances.

A newborn's first cry is a demand for control over its needs, which are comfort and food initially. From then on, a parent's job is to meet the baby's further demands for control in concert with the parent's willingness to give in to certain of those demands and to counter others, according to personal necessities of life—like sleep—and cultural directives, such as not being too permissive and spoiling the child.

Parenthood truly can be defined as a contest of wills for control over the child's relationship with the parents, which then affects individual relationships with other family members and society as a whole. Entire fields of scientific study, experts, and folklore exist to hopefully aid the parent with managing the child's attempts to control his or her environment, including food preference, bedtime, education, sexual topics, and clothing styles.

We can all chuckle at the truth of what raising children means and easily agree that the issue of control is front and center, and very obvious. But the matter goes further than this and has a greater significance; although parenthood is a familiar topic, it's nonetheless an example of what all life entails: the attempt to control an organism's environment.

It's a simple idea that is universally applicable to all individual life forms. Like a human child, a worm seeks to control its sources of sustenance and eliminate threats to its control of comfort, protection, sexual stimulus, and other facets of its life. Assuredly, the worm functions almost exclusively upon genetic directives, whereas the human child has a more complex mixture of genes and learning but even the worm can learn to redirect its behavior because of obstacles to control of its environment, as demonstrated in simple laboratory experiments and observations in natural settings.

Again, the point is that all life forms share this one characteristic that directs our behavior: The motivation for individual behavior is the attempt to control our environment. But, isn't that rather obvious, one could conclude? Of course, we try to control everything around us, our environment; and, logically, other living organisms do too. So what? How does that make a difference in anything? Yes, it's obvious. However, perhaps since it's easily assumed to be true, the concept hasn't been examined more closely to determine the extent of its implications.

Take any living organism and examine its behavior viewed from the perspective of control, as I briefly did with a human child and a generic worm. Is there any aspect of its behavior that isn't motivated by an attempt to control some elements of its environment, including not only the physical environment but also its relationship with other living organisms?

A worm reacts not only to temperature, humidity, and the particulate substrate in which it lives but also to other worms of the same species and predators and resource competitors of different species. A human child not only reacts to its physical comfort but also the parent as a source of food, reassuring hugs, and protection from nightmares and siblings—and also big shaggy dogs that play and bees that sting. All are part of the child's environment, and it tries to control each as best it can through smiling, crying, laughing, pushing, hugging, and later through language as a more efficient means of engaging in attempts at control.

Obviously, not all efforts are successful, and learning is the adaptation of behavior in order to refine control. A worm learns where to go to get the most food and where to best avoid predators. A child learns the limits of crying and the efficiency of language. The truth is that one, the motivation for learning behavior occurs as the result of an attempt to control some part of an individual organism's environment, and two, this holds for all life forms.

27

Food and Nutrition

At the most basic level of life, genetic instructions stimulate control-oriented behavior. For our worm, most of its behavior is of a nutritional, reproductive, or protective nature. Genetic code dictates chemical balances that need to be met and maintained, such as specific nutrients, hormones for sex, and temperature requirements for cell integrity. To achieve these, the worm must control its environment as much as possible, hunting for food, seeking a sex mate, and looking for a suitable place to reside. Beyond this, its need for control is minimal.

A person, like a worm, has genetic directives that create similar demands to be met. A human individual, however, has much more complicated behavior involved in controlling its environment. Take nutrition, for example. Genetic codes in our body cells require that specific minimum levels of nutritional elements are needed for the cells to function; without that, the cells die and the human individual dies. Therefore, a person seeks food to satisfy the dictates of his genes, but how the individual controls which kind of food is obtained is complex.

Initially, a baby crying for breast milk is a simple control behavior. After that, control becomes largely a matter of rejecting some foods offered by the parents and accepting others, and the offerings by the parents are set by their control of what they learned to accept from personal experience and by their cultural norms. In one society, offering varieties of pureed vegetables from small jars may be the choices a child has for acceptance; in others, pounded manioc, rice porridge, or mashed beans and potatoes served in communal bowls may be the options. As we get older, the boundaries of our choices become broader, or more restrictive, based especially upon our financial circumstance of being able to afford more or less of what we like and need.

Yet, there are also medical reasons for developing our food selections, such as nutritional and other genetically driven requirements. Some diseases also force individuals to eat a specific diet, for temporary periods and often a lifetime. Additionally, exposure to different cultural norms of cuisine because of marriage, change of residence, or travel can also change an individual's food preferences. Regardless of the source of stimulation, it all comes down to the fact that the motivation for eating and what humans eat is control, a combination of biologically driven necessity and conscious decision.

By extension, so it is with all human behavior. What is true for eating is also true for sexual attraction, clothing choice, defense against aggression, and kinship affiliation, to name a few behaviors. For some activities, the imperatives of genetics are strong initial motivators, with cultural imperatives subsequently molding the specific control which the individual chooses, as in building a shelter from the elements. With other behavior, cultural imperatives almost exclusively influence the individual's choices for control with little or no genetic inducement, such as in physical and vocal expressions of beauty. Regardless of the motivating factor, whether genetic, cultural, or a combination, control is still the motivating factor for these.

A Human Control Scenario

Taking the example of food, it's instructive to consider the utility there is in understanding that control is a defining factor in human behavior. When there's a problem that involves food and child-rearing, two questions can be posed: one, who is involved in attempting to control, and two, what is the control about? Using the familiar situation of getting a child to eat a particular food item, we can first establish that the parent wants to control what the child eats, and the child wants to control what he or she wants. This seems fairly simple.

Next, we establish that the subject of control is about some brand of cereal that the parent terms "junk food"—Brand X. The parent says, "It isn't good for you," and the child says, "It tastes good; " and the parent replies with statements about nutrition, and the child replies with the endorsement of some cartoon superhero, both sides subtly bringing in outside control agents to support their position. But is this contest truly just about cereal? Clearly, the basic motivation for eating is biological, but after that, it's a matter of culture and choice. The apparent questions are whose choice and what is it about.

Yet, an additional third question as to why control is in contention here is also very important because asking just who and what is only part of the phenomenon of control: Control happens for a reason. In this example, nutrition or taste is seldom the sole issue; we must look more deeply and ask what is to be accomplished by the parent or the child getting their way. How does the resolution of this control situation fit into the total pattern of control relationships between parent and child? What other influences are supporting the motivation of each in seeking control? Are there other people who are influencing each side? Are there involuntary, biological stimuli playing a part in the motivation of either or both parent and child, etc.? We can examine some of these possibilities with our example.

It may be that the child is reacting not to this particular parent's (say, the mother's) insistence that he eats a healthy cereal, but that he's reacting more to the fact that he was sent to bed last night without playing a favorite video game (due to inappropriate behavior earlier in the day) and that the mother was the one who enforced the decision both parents made. Moreover, the mother was tired from the day before because of particularly upsetting events at work and not getting the sleep she needed because the unhappy son kept making subtly disruptive noises until late at night. Besides that, he was wearing a shirt bearing a logo she had complained about previously, didn't put on the freshly washed pants she had laid on his bed, and had purposely gotten up late on a school day. Moreover, the son was still unhappy that his father had missed a soccer game the day before and his mother never noticed the small ceramic decoration he had made in art class and left on the kitchen counter, half-hidden by some pots since he didn't want to be too obvious in seeking attention for his efforts.

It would seem to be clear that Brand X was not the real focus of conflict in this situation. Asking why there's a contention of control promotes a broadened examination of the circumstances surrounding the conflict: that is, the goals which each actor seeks and who has an interest in control of some part of the actor's behavior. The mother was suffering from biological directives because she was tired and was bringing to bear previous control events and cultural issues of what a good parent should do. The child also had psychological issues of emotional stability involved, as well as the memory of previous control issues and cultural expectations of family values. The father was a player too since by missing the soccer game he also provided another layer of control and value issues that were brought into the situation. In sum, it was the familiar family situation of "one last straw" that really moved the control issue to a head: Brand X only provided the point of contention, though the mother most likely did feel that it truly was junk food and the son most likely really did think it tasted good.

Undoubtedly, there are other elements that can be brought into play for this example, especially the history of family events over the years where expectations and conflict were involved. But that only further reinforces appreciation of the complexity that is involved in human relationships, and accordingly, the complexity of control behavior. Examining the control elements of behavior can allow a clearer picture of what is happening, who is involved, and why.

Mate, Marriage, and Sex

Seeking a mate and establishing a long-term relationship is one of the most complex and universal behaviors of humankind. It has its origin in the genetic makeup of a person: stimulation of biological urges for behavior which enhances reproduction of the species. This isn't to say that the goal of sexual behavior is always to reproduce.

On the contrary, most human sexually related behavior has only the conscious goal of experiencing the emotional response associated with potentially reproductive motivations; reproduction itself is most frequently a tangential or unintended result of sexual activity.

If this were not true, then the culmination of every instance of the sex act would lead to biological processes of reproduction. Obviously, this isn't the case; otherwise, the world population would be a more staggering number, accompanied by all the consequences, and everything from an innocent kiss to intriguing clothing styles to most of our entertainment media wouldn't exist. So, it's safe to say that reproduction isn't the goal of all human sexual behavior, though its origins are undeniably biological.

At some point in their lives, every human begins to have the physical urge to respond to the biological stimulation of their sex organs, and it's something they have to control. How that's controlled is determined by one's culture, prior life experiences, ability to make decisions, and the type and strength of the biological urge. Hormones (testosterone and estrogen) and genetic orientation (male, female, transgender) largely control the type and strength of the sexual motivation. A person has little control over that unless other chemical or physical alterations are used to change those parameters.

However, an individual's culture provides a plethora of behavioral control guidelines for expressing sexual behavior: for example, strict religious beliefs in contrast to the freedoms advocated by popular consumer advertising worldwide. The specific cultural choices of activity which the individual makes will develop a history of personal experiences, positive and negative, to provide a measure by which to make future choices. At all times, for every decision, there's a constant interaction between biological stimuli, cultural guidelines, and experience that an individual reacts to in attempting to control one's sexual actions.

In trying to understand the specific behavior of an individual, then, asking questions about control can help. For example, a teenage girl choosing to wear "revealing" clothing that's "beyond her age" is trying to control the image she thinks is best for her, given the exposure she has had to commercial advertisements and peer pressure. She especially wants to control perceptions of her by other teenage females and males and attract them into her social realm of control. The parents, however, also want to control their daughter's image according to what they think is appropriate, based on their cultural guidelines and experience.

To resolve this conflict, each side needs to understand what the other wants to control and then reach an agreement over who has how much control over what. Parents understanding her peer pressure, and teenager understanding the parents' own experience with it. Parents giving in some on her wanting to fit a popular image, and teenager giving in on the most revealing part of her choice of clothing.

Understood from the perspective of control, both parents and teenagers can accommodate the other's viewpoint without getting into the emotionally based, sweeping accusations of "being mean," "never understanding," "being selfish," "looking like a slut," and so on. Of course, as suggested in a previous example, there may be a history of recent control issues on other topics relevant to the current argument about clothing, but they too can be separated out and dealt with at another time to keep the true nature of the clothing control argument clearer and more easily resolved.

There are many other examples of control behavior related to sex. For instance, rape is clearly about control and not necessarily having the sex act as its main goal. For example, it may be displacement of aggression towards a person who reminds them of someone who, in the past, molested the perpetrator; it may be a way of controlling one's feelings of confidence about women in general, by keeping them in a role traditionally defined in one's culture; it may also be, as in the case of warfare, a means of shaming the women having a particular culture to demean the status of the men in that society. One way or the other, the focus is on control.

More broadly examining sexuality, because of the mammalian genetic directive that females give live birth (except for platypus and echidna) and have an instinctive motivation to take care of the young, females have an inherent motivation to seek social interaction. They are interested in establishing relationships between individuals and forming groups based on cooperation. The male role in most mammal species, however, is generally different. Since they don't give birth and have a role primarily in conception, they don't necessarily have a genetically motivated protective control tie to the young. However, it should be remembered that the degrees of motivation to follow genetic directives do vary between individuals of both male and female sex according to their balance of testosterone and estrogen output, and this will strongly affect the instinctual behavior of both sexes.

Beyond these considerations, the male can still have a control tie to the mother and her young, since males were also once cared for by a mother, have a genetic basis for suckling and therefore physical attraction, and can also have a learned emotional bond to her and any siblings. The resulting behavior of that attachment can then be transferred as a learned incentive for finding a mate and to any resulting offspring; and control of those attachments becomes valued, especially as associated survival functions are included. The family unit then becomes something for the male to control and protect and a basis for extension into broader kinship units. Kinship units can additionally become the basis for further extensions of control for many social groupings and purposes, particularly in the human species, but variations of this are true for other mammals.

In summary, finding a mate is the basis for controlling sexual behavior in all mammals and has consequential genetic and learned behaviors that help define the character of specific species. Concerning the human species, there additionally are cultural rules regarding what is acceptable and what isn't for both seeking and gaining a mate: some formalized as documented laws and others as informal social mores or customs. As long as an individual controls his or her behavior so it accords with these rules, those actions will be acceptable, and seeking a mate can be largely unrestrained. Crossing those rules, which are control boundaries, brings condemnations of unlawful and socially deviant behavior and corresponding consequences.

Rules and Law

Let's take the idea of law as another example of control. Developmentally over time, human law is the result of genetically based instinctive actions for control over individual needs, which were subsumed into group control for the benefit of the group as a whole. Individuals came to find that if they modified those boundaries somewhat, maybe giving up some free choice at times, they could benefit from the group effort at food procurement, protection from environmental elements, or access to a sex partner. This boundary change for the individual was transmitted from one generation to the next as part of culture via observed behavior and eventually language.

As these control boundaries became more complex, transmission in forms such as oral history, storytelling, and religious concepts provided more organized modes of transmitting the requirements of participation in that society. Eventually, these rules for behavior were written down when that form of language transmission was invented, and much of early written language consists of rules and records of how those were applied. More cognitive abstract thoughts came later and finally appeared as what we would now consider being formalized rule of law.

So what, basically, is the essence of written law today other than a description of the control boundaries of individuals and groups as recognized by the highest appropriate level of social control: national, state, county, and municipal governments on down to the smallest social unit?

Beyond this, the exact degrees of balance between individual and group boundaries and the moral values to which the laws adhere vary tremendously: democratic, religious, tribal, dictatorial, monarchical, etc. Historical and current examples regarding this balance of control abound and seem to be almost exclusively about the role of the higher levels of control, such as governments.

However, in spite of being the foundation of law and civilization, the individual is somewhat sidelined in importance, and the focus is on the social group as if it were something existing in and of itself. The larger and more complex the social group is, the less the individual is accorded recognition of his or her control boundaries.

A democratic system of government, for example, may give individuals the choice of someone to represent their control interests in the larger governing structure, but with the massive requirements of operating a very populous nation or state, individuals increasingly tend to feel their interests aren't well represented and inconsequential. This can lead them to eventually rebel against the governing institution and take the matter into their own hands, joining like-minded individuals in a smaller social group to reassert their control or acting as a lone terrorist, for example.

The bottom line is that the entire structure of law, written or informal, is a constant struggle between individuals and groups of individuals to expand their control boundaries. As such, to examine and understand the governing of social groups, it's useful to not lose sight of the individual in the huge mechanism of government and laws.

When we want to understand why something is happening, we must discover and examine specific control boundaries of ordinary citizens and especially those of powerful individuals. We then try to determine how they are particularly influencing and being influenced by the larger workings of interconnected social groups, other individuals, and the laws that enforce a society's control. Hence, the flourishing of political biographies, historical narratives, polls, and daily commentaries in the news media serving the purpose of elucidation for those individuals wishing to evaluate how the boundaries of the governing politicians match or don't match with theirs.

Social Groups

Rules and laws, then, are justifications for the control of individual members of a group. For humans, there are countless tomes of opinion and description dedicated to how these rules and laws are carried out in social groups, including governmental and political action, but our interest here is on how these groups fit the concept of control. Furthermore, we need to understand how our human groupings fit into the larger picture of groupings for all species.

Also, it must be reemphasized that the individual is the locus of control. The individual can exist alone but the group can't; without individuals, there's no group. A grouping is an observer's perceptual construct, assembled by genetically driven behavior such as mating and raising offspring and/or cognitive concepts such as culture and an individual's memory.

Depending on brain complexity and species, an individual perceives one or more other individuals and is genetically motivated to respond to them as a group, as a single herring fish would in a bait ball or an annual spawning event.

Or in the human species, an individual is genetically attracted to other humans but is also motivated by past personal experiences and cultural behavior. As an example, we use language and clothing style to respond in a group with appropriate behavior, as a member of a sports team or a theatrical company would in the proper performance venue.

This importance of the individual is key to understanding control regardless of the species. Since an individual's survival depends on the ability to control both the non-living environment and the living environment, so too does the survival of a group necessarily reflect those requisites of the individual.

Group survival results from the summed efforts of individual members to continuously perceive those environments, interpret them, and transmit that information to other members so that necessary adaptations and actions can be taken to establish the control borders that define the group and individuals.

This is true not only of social insects such as bees and termites but also polyps forming coral reefs and E. coli bacterium clusters; all form colonies of like species and have control borders which they constantly struggle to maintain. And of course, there's the human species which likewise follows this same path of individual and group effort to define their boundaries and ensure their continued existence.

So, what is different about human groups from our other shipmates on planet Earth? It would seem that two differences are paramount: one, the predominance of culture in guiding control boundaries and, two, the inordinate influence that human groups have in establishing control boundaries compared to other species.

First, while most animal and plant species rely mainly on genetically determined physiology to establish their control capabilities, many animals also use the ability to learn and have memory. Humans and some other sentient mammals such as cetaceans and other primates additionally have culture, but so far as we know, human culture is much more complex than that of these other culture bearers.

Through genetic evolution, human physiology has made this possible through the development of a very adaptable physique capable of flexible mobility, grasping hands, and a large and complexly structured brain. Therefore, humans have been able to develop a level of cognition essential to the formation of cultural concepts, the means for recording and transmitting those concepts, and the creation of artifacts that increase their value and significance. When individuals coordinate all these factors with each other in a group association, the result can be powerful enablement of individuals to enhance their control boundaries and have a better chance of survival and transmission of their genes to future generations.

Second, given the above advantage of possessing culture and group cooperation compared to other species of plants and animals, humans have had a very significant influence on the living environment of those other species (e.g., decreased biodiversity and population) and the non-living environment as well (e.g., mineral and water depletion) because of their efforts over hundreds of millennia to establish and maintain control boundaries.

This advantage of culture in developing complex social groups has multiplied the power of the individual to establish and benefit from the boundaries enforced by the group. Of course, some groups ultimately don't benefit because of warfare, for example, but overall, the effect of social grouping, and especially that achieved through the formation of advanced governmental systems, has resulted in tremendous environmental influence. The cultural advantage of organized individuals for coordinated labor and creative concepts is unrivaled in the planet's history.

While the impacts of early humans on the living environment have been scientifically documented, such as the deforestation of Western Europe for agriculture and the hunting to extinction of Ice Age megafauna in the Americas, the effect on the non-living environment has also been increasingly documented. This is especially true within recent times: for example, the unprecedented consumption of nonrenewable resources for habitation, manufacturing, and transportation as well as the production of specific consequential atmospheric gases. Culture has not only given humans a great strategy for their survival but also an outsized role, compared to the numbers and influence of other living beings, in determining the boundaries and existence of those other members of the planet. Significantly, the greatest asset in achieving this role has been social groups, especially those of economic and governmental structures and the concomitant political power to effectuate that control.

Governmental Structure, Political Power & Economy

In terms of control, a governmental structure is a framework by which individuals accede responsibility for establishing and maintaining specific control boundaries to a grouping of individuals, either through force, voluntarily, or a combination of both. Political power is the enforcement of cultural mores, customs, and laws that validate a group's behavior as deemed necessary to enforce a governmental structure for control boundaries. An economy is a society's control of material and food resources from both non-living and living environments for the production and distribution of products according to cultural patterns of behavior. Other definitions could be used but put in terms of control, these will do.

Governmental structure, political power, and economy function interdependently to fulfill the control needs of an individual. Take away the necessary bodily population of individuals, such as by disease, conquest, resettlement, or diaspora, and it all collapses. Take away the culture, such as through language change or eradication of speakers, major institutional changes of religious beliefs, insufficient technology to meet demands of climate or warfare, and lack of resources to maintain food, artifacts, and physical structures, and it all collapses. Of course, some of these will take a length of time to eventually transpire while others will be sudden, both of which have a multitude of examples provided by history.

It must be made clear that these three cultural concepts are just that—ideas thought up to organize social behavior. None have an existence beyond their utility for the individual. Therefore, the individual must be motivated to believe that a particular format of each is in the individual's best interest in aiding their control of circumstances, which results in recruitment efforts varying from persuasion (advertising, propaganda, and reward) to coercion (punishment, pain, and death).

Whether the format is representative government or dictatorship, right-wing or left-wing politics, or socialist or capitalist economy, each vies for individuals to adhere to particular cultural beliefs in those categories, by one means or another.

Thus, when trying to understand what all the cacophony is about when assaulted by attempts for persuasion regarding these, it's necessary to set the situation in proper perspective by utilizing the concept of control. Keep in mind that individuals, both those seeking to control and the ones being courted, are all contending for personal control beyond that of the social grouping; understanding what level of control, personal and/or group, is being sought will help clarify the conflict.

For example, if an effort is to be made to change or repel the control-seeking activity by members of an advocating group, then the strategy is obvious and quite simple. First, identify the individuals involved and their control motivations, as suggested: Determine what is in it for them, personally—beyond the group's benefit. Then, using cultural concepts (adaptations of those the instigator is using or new ones), the defending individuals must solicit or join members of like belief (convincing instigating members to change sides or recruit new members) and mount a counter-attack. In effect, fight fire with fire, control with control. Reduced to essentials, that's what politics is about. There isn't any other recourse if a change in the situation is to be made, since every life action is always about control. It's just good to rationally structure the situation in a simple format of reality that explains what's happening.

The gist of all this is that it takes individuals and their participation to meet the demands of control for that social group to be successful. This is true not only for the three organizing concepts currently discussed and any associated social groupings but also for all social groups, whether they are based on kinship, education, ethnicity, charity, or religion. They all work the same: based on cultural concepts and reliance on individual membership, without which they can't exist. This should be remembered and examined closely as a focus on studying what has happened in the history of human society and in projecting what may happen in the future.

Communication

All living beings rely upon communication to define and maintain their control boundaries. They are compelled to be aware of each other's presence and respond because of this basic principle of life. So, how do we communicate with each other to do this? The answer fundamentally lies with the capability of each species. All species communicate among themselves (intraspecies) and between species (interspecies) in various ways such as through sight, sound, odors, behaviors, and physical objects—just in different forms.

Think of the male Australian lyrebird which uses mimicry vocalizations, showy plumage, and a constructed dance platform to attract females for potential mating as an example of intraspecies communication to control some of its environmental needs (we can easily find interesting human equivalencies). Also, scent-marking of territories is well-documented for wolves, as is the sound delineation of territories by songbirds.

Taking plant life as an interspecies example, research increasingly shows how all plants constantly test their environment through chemicals released in the air and water to let other plants and animals know of their presence and to continually attempt to obtain water, minerals, and sun exposure to survive and thrive. For example, plants of different species in forests use a layering effect of growth-patterned stories to gain needed sunlight exposure or to crowd others out over time in a succession of species. To counter this interspecies conflict, many plant species communicate with members of their kind, often using other species for transmission of helpful information: a combination of intra and interspecies communication.

For example, many trees, particularly in dense forests such as birch, have broadly spreading root systems and an intermingled network of fungi species through which the older and more mature trees send out chemicals to recognize younger tree saplings of their species. Once verifying chemicals are sent back via the fungi, the mature trees divert essential nutrients to the younger ones to aid them in their competition with non-birch trees. On a smaller scale, home gardeners grow plants species which avoid each other in their growth patterns to best set their boundaries of control for accessing sunlight and nutrients, but there are also plants, such as most legumes, which have a mutualistic relationship with nitrogen-fixing bacteria and mycorrhizal fungi to provide that important nutrient for them.

Regarding human intraspecies communication, individuals employ language and facial clues, behavioral actions, structures like border fences, and physical objects such as guns or gifts to let other humans know of their presence, needs, and desires regarding boundaries of control. Other individuals, singly or in groups, then react with their boundary expectations, and the adaptation task begins. Examples of this process are war, marriage, education, economic exchange, neighborliness, and so forth—all stimulated by the need to control.

Of these, warfare is certainly the most forcefully overt intraspecies human assertion of control. Communication with the enemy is done by threatening language messages but also by posturing and use of weapons, visual images, economic disruption, hacked communication modes, social propaganda, and ethnic, racial, and religious identification symbols, to name a few. In addition, uniforms aren't just so soldiers will not mistakenly get shot by their buddies but are also to communicate identity and unity to one's own group, and to the opposition.

For communication with non-human species, virtually all our contact with other species of the living environment results from searching for food sources, habitation locations, and artifact material resources. In seeking these, we announce our presence through the same sight, sound, odor, behavior, and physical object mediums that other species use—just in different, more aggressive forms.

Consider the automobile and its accompanying paved roadways, service, and manufacturing plants, oil wells and refineries, air, sound, and light pollution, animal roadkill, mining extraction, buildings, and cities to house drivers and families, etc. All contingencies of that single cultural artifact have significant interspecies control communication messages: that this land and water is ours, only approved plant and animal species are allowed in designated areas, you will be killed if you trespass, domesticated animals must obey human rules and will be rewarded if complied with, etc. Contingencies from the automobile also have intraspecies messages for members of human society regarding government and law, economics, family function, migration patterns, leisure time activities, and so on: all from this single element of modern civilization.

In sum, as with other life forms, most human communication is a combination of intra and interspecies control efforts, but in our case, culture makes communication much more complex as it intertwines with all forms of behavior and its generated artifacts.

Interestingly, both types of communication are frequently meant to deceive or hide the efforts of individuals and groups setting boundaries of control. Military uniforms, vehicles, aircraft, and other battle gear are classically camouflaged in warfare, as are hunting clothing and related shooting and fishing gear for those humans seeking to stalk prey to eat or hang on a wall to show their prowess. The rest of the animal and plant species are no different in using camouflage, though theirs is primarily genetically driven and not from cognitive brain function via culture, as it mostly is with humans.

The octopus and cuttlefish are masters of instantaneous skin coloration and texture changes while hunting or to deceive those who hunt them. Salmonella bacteria hide within macrophages to disguise themselves to avoid detection by immune system cells during a bodily infection. Flowers mimic shapes and insect pheromones to attract bees to them and encourage pollination of the flower species.

Sundews and Venus flytraps are classic deceivers of diverse insects by the traps formed from their sticky and aromatic body parts. Butterfly larvae mimic the coloration patterns of toxic species to avoid attack by birds, and one of the main functions of the patterns and coloration of adult butterflies is to likewise warn off potential predators through mimicry of other animal species.

Many examples of similarly camouflaged human communication exist in the form of coded messaging, political propaganda, visual art styles, email phishing, viral infections of computer software, social media on the Internet, commercial advertising, and on and on. Camouflage for both beneficial and nefarious communication control purposes has a long and extremely interesting evolutionary history in bacteria, plants, and animals—including humans.

Communication Throughout Evolution

For all living species, communication has always been important in acquiring control of the physiological necessities of food, water, and shelter. For less complex animals, this has been primarily an individual task since they generally have no need for cooperation: simply find the necessities and take them, if possible. Disease bacteria, for example, just invade a host and exploit it; of course, there's some competition from others and control has to be established, as with any living being. However, it's a rather simple stimulus/response control scenario.

With more complex life forms, the process of communication has become more involved, especially with sexual reproduction involved. For example, most insect species must individually find the survival resources they need and mate once before they die a short time later.

However, in other species having multiple mating and longer lifespans, more social behavior is required, as with birds. Mammals require yet even more communication since there are more mating opportunities over years and they need more time to raise their young. Live birth of offspring and extended care have huge consequences that force extended boundaries of control for both the individual and any affiliated social group.

Taking a quite intelligent mammal for example, elephants are social animals that use control boundaries collectively determined through varied communication skills as they forage for food, use water, seek shelter, and raise their offspring. They strive to keep individuals compliant as members of kinship groups, particularly concerning the demands and advantages of sexual reproduction, to make it possible for the group to better survive together than as individuals alone. Of course, individuals ("rogues," primarily males) that forgo the ties of group control do exist, and do in most other social species too. However, the basic fact is that live birth and extended care force mammals, such as elephants, into greater social behavior and complexity of communication while pursuing control of their environments.

Now, viewed from the perspective of a human evolutionary time scale and the demands of mammalian physiology for social interaction, we can see how reliance on intraspecies communication and the relatively rapid evolution of language would be a natural outgrowth of early humans also seeking increased control over their environments. For mated couples, successful speech development over extensive time and generations would gradually surpass initial reliance on physical gestures and facial expressions.

Having better speech skills, couples could better communicate with their children and other families to eventually bond as a larger group which could more effectively control their living environment of plants and other animals and adapt to their physical environment—all as part of their overall capability for more complex cultural development.

From this point on, the results of advanced communication ability would be many: differentiation into disparate language groups which separate speakers from non-speakers and establish other social and cultural barriers; development of representative symbols for language and mathematics, which could be transmitted by means other than vocalization, such as writing and digitization; construction of physical structures such as buildings and monuments which symbolize and display the power of a particular social group in contrast to other groups; and so on throughout human history.

Human interspecies communication would grow out of the requirement of obtaining necessities of food, water, and shelter by controlling those elements of life. For a simple example, hunting involves contesting the control boundaries of other animals by killing their species members for food or raw materials for tools and clothing. Individuals of the hunted species then adjust their territorial limits to that of human hunting activity. Both hunting and gathering eventually lead to the domestication of useful animals and the cultivation of plants for agriculture, which allows large scale control of food sources for humans. Appropriating access to water and resources for shelter and materials would follow the same path of control boundary extension.

In sum, this brief comparison shows that humans share a common basis of communication behavior with other life forms, differing essentially in degrees of complexity developed over time and evolutionary processes. Because of our unique physiology of adaptability with grasping hands, flexible limbs, and tolerance for physical environmental variations, combined with our brain structure and more varied communication capability through culture, we have greater potential for extending our individual and group control boundaries to exceed that of most other social animals. We may not always make better decisions on setting control boundaries than they, but we definitely have the potential to do so.

III

DIALECTIC DIALOGUES: Looking Within

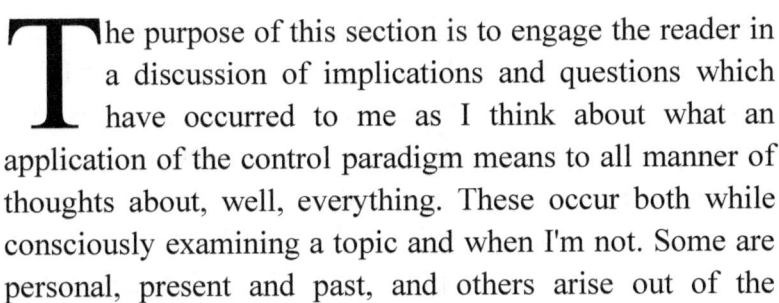

The purpose of this section is to engage the reader in a discussion of implications and questions which have occurred to me as I think about what an application of the control paradigm means to all manner of thoughts about, well, everything. These occur both while consciously examining a topic and when I'm not. Some are personal, present and past, and others arise out of the commentary and writings of others, present and past.

As with any dialogue, the discussion moves from one idea to another related idea. I've tried to bring up implications of and formulate tentative conclusions on some topics and also raised unanswered questions on those same and other topics to prod further thinking for the reader and myself as well.

All are attempts to explore examples of control within the human mind that are used to explain what is observed and thought of when encountering other living beings and elements of the inanimate environment. Our perceptions demand an explanation, and we consequently try to provide that control. Summaries of discussions about the initial concepts are provided later in this section.

Memory

If we accept that rational decision-making as a tool for control isn't exclusive to humans but is shared in various levels of complexity with other organisms, then what does this tell us regarding the several human concepts of self, an idea that has been the subject of philosophical and religious discourse for millennia? As a start, it might be useful to pursue the viewpoint of memory being the same as or a major part of the self, spirit, soul, or consciousness of an individual life. Is memory actually what is meant by these words given that memory is a unique mark of an individual and these concepts do imply a special individual uniqueness as well?

There is verifiable scientific evidence of the actual physical basis for memory via chemical and electrical processes in the brain, particularly as obtained through fMRI and other scanning of the brain where various cognitive activities can be detected, including memory processing. So, something physical is definitely involved with the concept of memory aside from philosophical and psychological lines of thought. But if memory is the same as a self/spirit and is evidenced by physiological processes, then it would follow that if the body dies, the brain memory processes would stop, and therefore, logically, any self/spirit would also cease to exist.

Given the implications of that reasoning, it may be useful to look further into the relationship of selfness concepts and memory to evaluate the conclusion just stated. To begin, it's possible to think of memory as basically the personal history of an individual being's control activities throughout their life. Memory is what most people consider to be a primary attribute of being a human being and is what we think would be missed when we consider death and attachments to those memories—especially people, pets, homes, and places we've been.

So, experimental studies of both human subjects and other species can be made. The results are then compared to how memories are formed and processed concerning the complexity required for controlling their environments in daily living activities. By clarifying whether memory is a universal feature of all or some life forms and what its parameters are, this information would aid in assessing if memory is actually what selfness concepts such as spirit and soul refer to.

However, rather than using only comparisons between humans and other animals in their fully functional cognitive status to assess the importance of memory, we can also take the experimental subtractive approach as is done with lab animals such as mice to determine what effects result from reducing cognitive capacity through surgery or chemical injection for medical research. Much has been learned about memory by this method. However, such a subtractive approach isn't morally acceptable with human subjects; but in fact, the opportunity for such study is available when memory is impaired due to unintended causation, as with traumatic brain injury, stroke, or dementia.

As I can attest from a family member's subjection to the ravages of Alzheimer's dementia, the role of memory in a person's control capability is all-pervasive. Not only are there the initially minor inconveniences of forgetfulness which steadily increase to dangerous levels of impairment, as with household safety and wandering, but also the failure of communication.

Taking just this example of communication, to speak meaningful sentences, one has to be able to remember what was intended to be said immediately in the next few words but also what had already been said, since past words are a guide to the use of future words.

Hence, Alzheimer's patients increasingly get lost in where their conversation is going, eventually being unable to complete a sentence, and finally being unable to speak words but only unintelligible sounds, if even that much. Memory functions to retain the meaning of words and their combined ideas, and also providing the means for sharing those ideas through sentence structure in spoken forms, and a similar problem occurs with written forms of communication.

Of course, communication is only a fraction of the results from taking memory away. As different areas of the brain are affected, also lost are inhibitions, recognition of friends and family, lifetime events, location awareness, ability to carry out normal activities of daily living, and other aspects where memory is essential—which is just about everything.

These are all parts of a person's particular uniqueness that distinguishes them from other individuals: those elements which can be variously termed one's self, personality, spirit, soul, or consciousness. When that uniqueness disappears over time, it's very clear to others who share a lifetime of memories that the essence of the person is no longer there, in that body.

Who that person was, no longer is, but has changed and eventually disappears completely with death, except for, what? Anything? Hence, the hope by loved ones for some continuing essence of the person as a separate entity, not tied to a physical body. Perhaps some additional perspective would be informative.

When someone dies suddenly, the mourner is left with a paradox that is strongly infused with emotion. The person was just there moments ago, body and personality/self combined into one unit, and suddenly only a body is left. Where did that selfness go? The contrast can be overwhelming, and logic isn't very emotionally reassuring at the time, but some sense of incongruity, between what was just there and now isn't, seems to demand an explanation.

The feeling is that the person couldn't just disappear completely; there must be something left.

At this point, cultural institutions and beliefs can be comforting in providing explanations of the soul, spirit, and other concepts of a person's existence beyond the body. However, when a person's memory and affected personality/self isn't suddenly changed but is gradually altered over months and years, as in dementia, the observer is able to watch as all aspects of that person are gradually subtracted from the essence of his or her being until nothing is left but a body.

Overall then, how is the mourner of a death left to understand where the selfness went? In the case of sudden death, the thought is that a separate self/soul has always existed and is now suddenly separated from the body. As suggested, there is a multitude of cultural, religious, and philosophical concepts that afford individuals a control structure for dealing with their emotional and reasoned responses to sudden death. With a gradual death such as dementia, the selfness issue becomes a bit more complicated.

Is it that there are parts of a once complete soul or self that go somewhere on a daily bit-by-bit process over the years of dementia degradation until it's reconstructed somewhere "out there?" And what is it that remains in the body while the transfer is being made piecemeal—a separate soul or self that gradually diminishes, plus the reconstructing one? Alternatively, maybe the disease is just increasingly hiding the soul or self until the physical being ceases to function, and then the separate entity is revealed elsewhere intact? Given that the two scenarios of sudden death and gradual deteriorating death both end up the same way, with only a non-functioning body left, it seems difficult to explain how one path slowly regenerates or reveals a soul and the other path does it instantly fully formed. Yet, I suppose some cultural explanations for this have also been formulated.

Nevertheless, a much simpler and more consistent explanation would be that an individual's history of setting control boundaries—cognitive memory—constitutes what is termed self, soul, spirit, etc. Accordingly, as the individual becomes unable to maintain the brain cells where memory is potentiated due to, say, slow dementia or sudden heart attack, the body can no longer function as before, the observable behavior of that body changes, and control is completely lost.

Regardless of how long it takes, no memory means no more self, soul, personality, and so forth. There's just nothing left but a lot of resultant unrelated disintegrating molecules. The only evidence of the prior existence of that individual's self, however it's termed, now resides solely in the cognitive memories of other individuals or in some cultural form of those memories—but only for as long as they continue to be maintained.

In any case, the individual's original is still gone forever; there's no separate selfness for humans, nor any other life forms. Yet, broadening the perspective a bit, there are ways for some types of memories to continue and be used as tools by members of a species to effect control of their environments.

Three Ways of Passing Something on After Death

First, biological matter is passed on as a physical living part of an individual by actual molecules carrying DNA genetic coding being copied and transferred from one person to another during reproduction. The same would hold for any living being, whether sexual or asexual. This physical coding, as genetic memory, could continue in replicated form for generations long after an individual's death. Of course, with evolutionary selection, coding sections will be lost or changed over time, but a large portion would continue for some duration, After all, every living thing today shares a minimum of coded material reproduced from our original common ancestors and uses genetic memory in controlling survival.

Second, an individual's cognitive memory (specific lifetime memories of events, thoughts, conversations, etc.) could be passed on orally for a few generations as long as it's shared with other individuals; it would not be the exact cognitive memories made by the individual but would be a recollection and interpretation of those memories made by others. Being shared with many other persons, this information would develop a commonality of memory about that person to be passed from one individual to another, spread via oral communication.

The major difference here in contrast to biological memory transmission is that the memory does not have to be held by a descendant—anyone within the needed immediate proximity of personal experience can have the memory in their brains, including friends, casual acquaintances, and sharp-eared gossip mongers. In addition, it's more short-lived since once memory holders die, that memory reference disappears, resulting in the eventual eradication of all cognitive memories about an individual in a short time. This would be true as well for any non-human animal species which might be capable of some level of oral language.

Third, culture is also basically a type of memory resulting from the shared cognitive memories of individuals (largely anonymously) and offering a unique combination of some memory features that the other two transmission processes separately have. For one, it too can be quite long-lived over time in being able to transmit information about deceased humans, individually and collectively, through books, electronic media, stone writings and monuments, roads, and buildings: all the usual information resources of history and archaeology. For another, it also does not have to be transmitted in lineal descent from one person to another directly; the recipients of the information don't have to be related nor have proximal contact, which allows a potentially wide distribution of that information over time.

Moreover, culture does not have to be physical material but can be behavior that is transmitted as memory. If accompanied by material-based instructions such as in durable media and decoded, cultural behavior can be reenacted; otherwise, the behavior must be taught and transmitted orally from generation to generation, though not necessarily through related generations.

Whereas material culture is more restricted to humans, behavioral culture may be widely transmitted through some span of time between individuals of several other species, for example by birds, other primates, cetaceans, and elephants. Also keep in mind that although culture is a type of memory, it's primarily the result of genetic evolution that made the physiological potential for cognitive memory possible.

All three memory types are instrumental in an individual's ability to control their living and non-living environments. While all life forms must use the first, the second and third are additionally incorporated to varying degrees into other species' survival toolboxes of control techniques, depending on their evolutionary level of development. Besides, all types are shared from one generation to the next with genetic memory being the most enduring over time, cognitive memory being the shortest, and cultural being able to last with a wide variation in time duration. Moreover, unique individual genetic and cognitive memories are contained within the individual's body and brain, whereas cultural memory is essentially a social phenomenon resource—a sort of group brain—which the individual refers to when the limited amount of cultural knowledge stored in their brain isn't sufficient to aid in control situations.

Memory, Self, and Existence Beyond

With the previous discussions in mind, we need to utilize the observations just made as we now consider more closely how they relate to clarifying the concepts of self and existence beyond the body, and whether such is possible. This means delving more into the physiological basis for variations in memory production and how that relates to self.

For this further perspective, it must be emphasized that living beings need memory for controlling their survival, otherwise the same destructive actions will be taken by them again and again to the demise of the individual. Memory facilitates survival and consequent evolutionary adaptation. This adaptation is first set in genetic DNA coding as a form of memory so cells will replicate again and again successfully. Genetic memory is the foundation for the formation of all memory in all kinds of living beings, and the exclusive type of memory for less complex animals and plants.

As life forms became more complex, with animals, in particular, the brain and associated neurological network of cells developed the capability of recording positive and negative actions by the animate being (cognitive memory) which could be referred to instantly and, in concert with genetic memory, make decisions (i.e., learning). This cognitive memory coding in the brain exists only as long as the central nervous system is functioning because the individual is alive, in contrast to genetic memory which is coding set in DNA of the cell structure itself in all cells throughout the whole body and potentially continues beyond an individual's death through reproduction.

The evolutionary development of reproduction itself has played a role in the genetic memory of animals and plants. Beyond the simple replacement of existing cells of a living organism, the reproduction of entirely separate new individuals is more complex.

As the most rudimentary type, asexual reproduction is the generation of genetically identical offspring by such means as budding, binary fission, and fragmentation by one parent organism alone. The asexually reproduced offspring then have the identical genetic memory of that parent, except for infrequent random changes in cell replacement, which could potentially make small differences in the offspring's genetic memory (DNA) over time.

This type of reproduction occurs with simpler organisms, and a spirit/soul could seemingly be duplicated along with the replication of genetic memory. However, this would mean that a spirit/soul would also be virtually the same for each new individual; there would be no unique individual self, which is the foundation of the idea for a spirit/soul. Besides, this would only apply to life forms such as bacteria, fungi, and non-flowering plants that rely on tubers and roots, as well as many worms, yeasts, hydra, algae, and echinoderms. More complex life forms such as fish, reptiles, and mammals would be left out of a spirit/soul opportunity via asexual reproduction. Since there would logically be no difference between genetic memory and a spirit/soul in asexual reproduction, it would only be a false ideational difference in words. Hence, the existence of a unique individual self at this level is groundless.

In contrast, sexual reproduction is the basis for the procreation of all more complex organisms and results from two individuals with genetically different information, male and female, being combined to produce genetically unique individual offspring. In that sense, a sexually reproduced being could have the genetic memory of the parents. Yet, the complete individual uniqueness of the parents isn't transmitted in genes via the reproductive process since the two different genetic gametes (i.e., genetic DNA memories) combine to produce unique offspring, leaving out some parental features and keeping others, and not exact duplicates.

So, neither asexual nor sexual reproduction has a role in the possible production and transfer of a distinct spirit/soul from generation to generation into a different physical body as genetic memory is passed. Asexual reproduction would only generate identical soul/spirits, not unique ones; and sexual reproduction would generate a unique soul/spirit but would be non-transferable to the next generation because the parent genetic material contributed is always different with each individual produced. The new individual would already have its own distinct soul/spirit, with no place for mixed soul/spirits or ones that rotate appearance—not the usual definition of the concept.

Consequently, there is no reincarnation opportunity in either situation. Nor is there a possible unique spirit/soul existing separate from an individual's body anyway, as concluded from the earlier discussion on that topic regarding memory. Yet, with sexual reproduction and increased cognitive ability of more complex animals, the offspring's accumulated cognitive memory combined with a mix of parental genetic memory (e.g., physical likenesses) does produce a uniqueness of an individual which might be extended to the idea of a spirit/soul of an individual.

This is what's involved when we say something to the effect that daughters or sons are like their parents or some other relative, in physical appearance and/or their behavior. However, since only partial genetic memory is transferred via DNA, any similarity of physical appearance would have to be attributed to that: some shared DNA, by chance. In like fashion, the only basis for behavioral similarities would have to come from the cognitive memories of those who knew and remember the relatives being compared: They just notice that some actions are similar, existing in the beholder's eye. It would not be from any spirit or soul that was somehow transferred and involved in producing those similarities, neither as an entity existing separate from the physical body nor as reincarnated wholly into another individual living being.

Accordingly, when an individual dies and all genetic and cognitive memory is gone, what's left? Again, it would seem to be that only the chemical constituents of the body remain. Still, there imaginably could be something else left since cognitive memories about the deceased individual can continue to exist in other living beings for a short time as part of their memories.

Additionally, since various chemicals from the non-living environment are continually used for cell reproduction and maintenance, it's at least conceivable that some of these chemicals from the remains of known formerly living beings could be taken in by chance via food and/or breathing, and then become part of another being—such as a family member—who has memories of the deceased. Of course, the potential ingestion of chemical particles would have to be deliberate or resulting simply from proximity to the deceased.

Therefore, do those former individuals now "live again" in yet another individual, reincarnated into their bodily constituency? In some metaphorical sense that could be said, but would that other individual have any actual memory of past lives other than that of their personal memory of someone contemporaneously known? No, because cognitive memory does not exist in the cells of the brain and their constituent chemical material (which might be ingested) but only in the code processing functioning of the brain while the memory holder is alive. And that memory can only temporarily continue via cultural means of behavior and communication and through personal remembrances. There's no transfer of actual memory possible.

Granted, it can be psychologically distressing to contemplate losing one's memories, one's sense of self, but that's the way life is. As suggested, one's chemical elements conceivably could contribute to the continuation of many other lives and spread throughout the Earth and eventually throughout the universe, given enough time. Since matter can't be destroyed, then those non-living elemental remains will exist forever. For many persons, that can be a rather cold emotional consolation, though, regardless of the expansive imagery.

Broader Questions

But perhaps it should be asked if there's more than cognitive memory that makes up the self? Is everything past? What about present activity as a part of self? Genetic instructions for bodily functions/activities that direct actions of the body and consequent cognitive memories of those actions (such as a response to illness or a tasty meal), could also be part of one's self—as it takes place, as a person is in the act. But as soon as it happens, it becomes past and part of memory.

Therefore, for a fleeting moment, our self is memory plus what is happening at the moment. That addition of a temporal dimension to self would be like every second of existence being a mini-death because it becomes part of memory and the past and thereby relegated to the same temporality of all memory. Self would be a past streaming out from a continually changing present—much like a comet in outer space. And is there a place for the future? A future component to self?

It would seem to be only in the sense of memories and current behavior patterns being a potential guide for future actions, and resulting memory. So then, what is the present? Does a complete self exist only in that instant of time between memory and the future? Perhaps the word "personality" has some utility here as a descriptor of what the self is in present continuous time before it becomes memory.

Now, if we, being an animal species, have memories, don't other animals as well? Then, don't they too have personalities, maybe a concept of a self? Our human brains have larger centers for memory but that is a quantitative thing. It's not that other animals don't have any memory centers at all; they are just smaller and qualitatively less able to have complicated memory and therefore thinking abilities. Yet, maybe they could still be capable of self-concepts, and personalities, like humans can.

What about self-awareness as being the defining distinction between humans and other species? The problem is that it's difficult to show that other animals do or don't have self-awareness, except perhaps dolphins, elephants, magpies, and chimpanzees as previously mentioned. But it does seem that memory is integral to self-awareness and is basically a control refinement that at least we humans have, and perhaps other animals have different cognitive control refinements, which we don't have.

From another perspective, if plants have no brain or nervous system but they communicate by chemicals, how different is it really from animals? They communicate more slowly but they still communicate; then, is the speed of transmission the difference? Don't animal nervous systems also use chemicals (albeit with synapses and neurons), with some added electrical connections in the brain to speed up response? What about an individual plant's self-identity or a memory—is it possible? Why might an animal have those and not a plant? In essence, isn't it all just a matter of the control of atoms?

Furthermore, is there such a thing as an individual anyway, since we are all integrally connected like a plant community (with all the bacteria, viruses, fungi, etc. that affect our body's function)? Maybe there's only an *us* and not an *I*—an individual plant or animal.

The concepts of self, soul/spirit, a separate otherness, reincarnation, memory, and time seemingly become mostly a contrivance of words: that it's a communication by-product of our evolutionary development of brain complexity rooted in using language and a cultural consequence of us seeking control of our environment.

Furthermore, it results from trying to explain it all by inventing words to fit specific situations and concepts beyond our ability to assign completely precise meaning. Yet, it may be informative to examine these concepts in more detail and from different perspectives.

Concepts of Self

Briefly, there are several words that humans use to generally connote the idea of there being something besides just the physical parts of our bodies. They all suggest there is some "selfness" that exists, of which we are or can be aware: that it's the essence of what it means to be human, and if there truly isn't such a thing or it goes away, as when we die, then what is "me" as a person?

Now, going back to one of the principles of control, that since all living beings share the same requirement of controlling the essentials of life and therefore the same basic physical construct of cells, DNA, proteins, and related chemical compositions, the question is: If humans have this self/spirit which is supposed to exist separate from physiology, then perhaps all life forms do too, and if not, then what is it that makes us different?

All living beings have genetic elements that direct them to carry out specific behaviors to achieve a level of control of their environment sufficient for survival. As some life forms evolutionarily grew more complex, their nervous systems did likewise so that decision-making to perform these behaviors consequentially became more complex: i.e., see food/grab food was replaced with risk assessment regarding which food to grab and when. In broad terms, it's a matter of making decisions based on alternative information in contrast to automatic instinctive stimulus/response action.

So, is it that we are just a life form able to use more complex thinking which differentiates humans from other life forms? Perhaps we can discern substantive differences through scientific procedure and experimentation, at least that's how humans have tried to discover it in the recent past.

Cognitive sciences, such as cognitive psychology and neuroscience, are the most direct attempts along these lines, trying to explain various behaviors that seem to be driven by hidden aspects of the brain through experimentation and clinical studies that measure observable responses of the nervous system. Consequently, we try experiments on humans and less complex animals and note the differences.

On a simple and more experimentally accessible level, one potential for comparison is in the realm of self-identity, such as in experiments done with dolphins, chimpanzees, magpies, and Asian elephants, all of which demonstrate an ability to recognize themselves in mirrors as distinct individuals from others of their species. Even so, that's only one sensory input for identity. Dogs, wolves, and many other mammals use scent marking to distinguish themselves from others, to define possession of territory, and to exclude others from what they claim as theirs. Vocalizations provide another sensory marker for the same purposes.

How different is this from common human activities of visual appraisal in a mirror, setting fences along property boundaries, or the roaring cheers between sports stadium spectators seated in appropriate locations to hopefully avoid conflict between team supporters? While there are many more experimental studies with varying cross-species designs that have been done, it's doubtful that examination of comparisons with less complex animals and would find significant conceptual differences and instead would find the opposite: the commonality that all these behaviors result from enacting the basic principles of control.

I can't help but include a personal experience that seems to fairly exemplify the ability of another mammal to use awareness of self and simple reasoning to achieve verification of its existence and control. I was a child of eleven years old at a zoo, back in the days when moats did not separate visitors from exhibits, walking along a lion enclosure fence, and I approached an old male lion ambling half-awake along his side of the enclosure fence, going the same direction as me. I was intently studying him as we paced along close together when suddenly, with an overwhelming roar, he feigned a momentary charge directly at me and then instantly continued his slow, disinterested pacing—except for one instant of brief eye contact when he turned to cast a glance my way to let me know, terrified as I was, who was master and who was the prototypical meal material that he intimidated by choice. He definitely knew who he was, and so did I.

Of course, it can be argued that the lion was just making an instinctive action toward an intruding food source (me); however, it isn't that simple. The action had certainly taken the form of instinctive aggression, and he had surely learned through experience that he could get a spontaneous response from bothersome humans like me. But the key point here is that he was capable of thinking with reason to decide whether to pull his trick, as well as when and how scarily it could be done. He had learned how to use instinctual behavior to control his environment as best he could, and I had learned about my control boundaries as well.

Given this simple incident and many more formally set experiments, it can be said that at least some less complex beings do have an extent of awareness or consciousness of themselves and the ability to apply a basic level of logical decision-making to stimulus and response. Do humans have any level of logic beyond this? Of course, but it would appear to be a matter of degree in complexity that is applied to control the individual's environment and not any capability missing in all other life forms. Just like other beings, humans have a multitude of instinctive behaviors, and it involves identifying them and understanding how they are adapted to individual control necessities through learning to distinguish what is different regarding our species compared to others.

Consciousness

So then, what is consciousness of the human individual, since it appears to give rise to these cultural concepts involving selfness? Is there truly such a thing? If we are all a community of cells, tiny arthropods, fungi, and bacteria, then how do death and consciousness relate? We assume that death would be the end of consciousness, as with any cognitive process, and consequently the end of our community self.

However, our awareness of self largely ignores our community aspect, except for health issues, and we do not mourn the death of those parts of our broader community that cause problems. The selfness we focus on, which we term consciousness, could be most simply viewed as a series of control events recorded in our brain memory. Having a larger brain allows more control events to be recorded, and the feedback loop of memories gives a sense of consciousness. Viewed from an evolutionary standpoint regarding other animals, then, a small brain would mean less recording and less sense of self and consciousness.

And what would be the purpose of consciousness? Perhaps so more control can be exerted to keep the organism community alive, like all other control processes. Therefore, increased consciousness or self-awareness throughout evolution may just be another means of control, especially for more complex animals. So, when an individual loses control and cohesion, and its atoms disperse, consciousness disappears along with self since it's essentially not an entity but another fundamental element of control necessary for life to exist.

Revisiting the basic principles of control may clarify these points. Because of the requirement for life that all beings must attempt to control as much of their inanimate and living environmental components, that attempt to control requires all life forms to be able to sensorially perceive what is in those environments: that is, an information processing system of input from the senses and faculties for effecting behavior dictated by that processing system. Without such basic components, the life form is unsustainable; in fact, it never could have existed initially.

Evolutionary development of these components made every life form possible and has continued through time to winnow out those life forms which have unsuccessful variations of those components. If sufficient adaptations to either the inanimate or living environments don't result in successful boundaries of control by the individual, the individual being dies; and without the individual, there are no progeny of that variation (species extinction).

The relevant point here is the information system component. All living beings have one, but they are different for each species of living being. In addition (focusing on just animals for the moment since that's our best familiarity), each species has a variety of sensory inputs but none have exactly the same nor utilize those inputs the same. I would venture to say that humans rely on sight most of all, with hearing, scent, taste, and touch more supplemental.

Obviously, this isn't the same for all other animal species. Naked mole rats, in their burrows, rely significantly more on scent and touch than any other; in fact, they are essentially blind, as are many other animal species of insects, spiders, and fish that live exclusively in cave habitats. Bats, of course, rely heavily on hearing for their ultrasonic navigation and prey location. Ants rely upon scent and touch for communication with each other regarding food sourcing, outside predator defensive action, care of community egg hatching, and stimulation of the queen to produce her eggs.

Even in the realm of sight, humans perceive only a specific range of the color spectrum, whereas many birds and insects see various wavelengths of that spectrum plus ultraviolet which enhances details of colorful plumage and flower display invisible to us but of significant importance to them. Mantis shrimp have twelve color receptors (humans have three) which allow them to be more energy efficient by being able to distinguish potential prey having specific coloration, and the most nutrition, from those who don't have it—and thereby find the best meal to eat. The animal examples of variation are quite endless, and this is a marvelous thing because it shows the flexibility of organisms in meeting the perceptual requirements for controlling environments.

Since all living forms differently perceive and react to their environments, where does consciousness come in, or self, personality, mind, and all of that? If humans have the same basic necessity of receiving and processing information as all other life forms and we call it consciousness for humans, then do we call this process consciousness for all other life forms too? Or is it just a matter of words—and history, of humans trying to separate themselves as somehow uniquely differentiated from "animals?"

So, exactly what is meant by consciousness? Three basic aspects of the term can be distinguished: (a) being aware/conscious of environmental stimuli originating from outside the individual (exteroception), (b) being aware/conscious of stimuli originating from inside the individual (interoception), and (c) being aware/conscious of the relative position one's body parts in space and the amount of effort to move them (proprioception).

Consciousness of the environment outside an individual seems quite straightforward and can easily be validated by physiological tests and measurements of an individual's receptivity to sensory input, such as in oral responses to visual and other sense test stimuli and medical determinations of fainting, coma, or anesthesia levels. Likewise, medical testing can also be done for determining consciousness of what is happening inside the individual such as with EKG heart and EEG brain monitoring, fMRI scanning, and ultrasound recording plus many psychological experiments regarding pain, smell, and gastric sensing.

Consciousness of body position relies substantially on kinesthetic sensing at body joints by the individual for scientists to design physiological and psychological experiments as well as for human rehabilitation training. Psychological procedures set up to test phenomena such as the phantom limb syndrome are quite informative and help refine the concept of consciousness. This means that there appears to be a definitive scientific examination of physiological and cognitive phenomena to validate the existence of consciousness as a general concept.

Incidentally, we also informally attempt to discover what is going on inside an individual by asking how that person "feels" (health-wise or emotionally) or what they are "thinking about": the inference being that the individual is capable of being conscious of what is happening to themselves internally. It seems that the use of the word "aware" may be more appropriate in this informal use than that of "conscious" since consciousness has more medical and physiological connotations.

The terms self-aware and self-conscious, however, are a problem in using consciousness, informally and technically. It's at this point where the concept creeps into more definition problems due to the use of "self" which can imply a separation between "other" (outside) and "self" (inside).

Outside is easily distinguished: whatever is outside the body's boundaries of skin, hair, and nails; but inside is vague, unless one applies the same anatomical distinction of whatever is inside the skin, hair, and nails to "self." But that isn't what we commonly imply with the words self, soul, spirit, or mind.

That means it's a word problem, a matter of definition because not all people mean the same thing as to what the words connote—and that varies between cultures and languages. As with the translation of words in all languages, it isn't a simple task; some words or concepts don't have directly translatable equivalents, or in many cases, even remotely similar ideas. If the languages are closely related to each other, as are Spanish and Italian, it's an easier job than if they are more distantly related or even completely unrelated, as with Navaho and Samoan.

So, how is it possible to determine what a concept means in any universal agreement if there are such difficulties in communication? Given this situation, it's a wonder that any human can communicate with any other at all. This is why names for objects and activities are always easier to translate than immaterial concepts, such as is being attempted here, and we are enticed to just forget the whole matter and ignore the problems brought up. Yet, perhaps it's still worthwhile to pursue clarity.

Summaries of Discussions

Consciousness and Self

Certainly, these two concepts are difficult to delineate because they are quite involved and lie between the more easily definable and observable concepts of memory and personality and the still more nebulous ones of soul and spirit. Yet, there's hope if we apply the basic principles of control.

Consciousness is an individual's perception of and reaction to the environments outside and/or inside the body, a phenomenon that lends itself well to medical and neuroscience research to observe and measure. To be conscious, a life form must at least react to its internal environment, or to both internal and external to be called fully conscious. If the individual does not respond to anything in either realm, it isn't conscious but dead, plainly put.

Furthermore, in terms of control, consciousness results from chemical and/or neurological activity involved in the individual's seeking control of its environments through a physiological information processing system component that all living beings have in some manner. It's a control process of sensory perception and interpretation which can be a simple genetic-based stimulus/response in elementary life forms or be more complex in those life forms with more multifaceted chemical and neurological signaling communication systems.

With humans and most likely some other primates, cetaceans, and a few other life forms, this advanced neurology and consequent thinking ability related to brain size has further led to the capacity to develop a sense of self-consciousness.

Self would denote consciousness about the control activities of an individual being, by that being, as defined by whatever is within and on the cellular barriers (e.g., skin) of the individual, in response to other life forms and the non-living environment. The self only exists with consciousness, and memory provides the information which makes consciousness and self-consciousness possible. Take away either part and self disappears.

Memory

The definition and concept of memory would appear to

be simple: Every individual, of necessity, has a genetic memory of DNA because of reproduction from a parent(s) and, for more complex species, a neurological record of thought and behavior recorded as cognitive memory while alive and actively setting control boundaries over time. Memory functions to guide an individual's control activity. That's essentially it.

Individual genetic memory can be deciphered through DNA decoding, and cognitive memory can be decoded by reviewing past and present behavior of individuals, as in decision-making about food procurement and mating.

With humans, it's additionally decoded through analysis of communication in the form of gestures, spoken, and written language besides artifacts of a more permanent nature, all as part of cultural memory.

As discussed at length previously, memory isn't a separate entity and ceases to exist with the death of the body, except for some genetic DNA that will become diluted in offspring over generations through evolution and sexual reproduction, and some very temporary cognitive memory held by other individuals and a society's cultural memory.

For most of us humans, few of these memory forms are in effect for any significant length of time, which also is the situation for individuals of all other life forms with which we share this planet. So, we're not alone in this respect.

Personality and Mind

The definition and concept of personality follow closely from an individual's behavior and communication while accumulating cognitive memory: It's the communication and behavior patterns that are displayed to others as an individual uniquely responds to the necessity of setting control boundaries. What you do is who you are, in the briefest terms. Since personality is tied to cognitive memory, the personality capability of individuals in a species increases with the brain complexity of each species. How that individual behaves in accordance with their cognitive memory in present time is the individual's personality, his or her self at that moment. Of course, personality is a continuing formative process; it's always changing, but the reason for that is the necessary adaptive changes to one's ability to control one's environment, inanimate and living. And with humans, the social living environment is of paramount importance and the primary place where control activity occurs.

Since personality is based on cognitive memory, it's therefore subject to the same temporality that cognitive memory is. However, we rarely mourn the loss of personality, especially since other selfness concepts loom larger in the overall focus of life, and throughout our lives, most of us welcome personality changes as a part of maturity. Of course, some personality changes are socially frowned upon or even outlawed, but even those are part of the overall control process of personality and fit neatly with the general control paradigm.

Regarding the mind, we can make similar general conclusions: The concept is a product of our consciousness and memory in setting control boundaries that reflect the functioning of our brain. It's just a shorthand, handy term to refer to one's thinking and consequences thereof, such as, "This is what was on my mind," and what others wonder about another's thinking, such as, "What do you have in mind?" in contrast to observations about one's physical body and behavior. Like personality, it's another useful communication expression referring to the broader concept of selfness, and is likewise temporal.

Spirit and Soul

The last general concept of selfness to be summarily defined is the idea of an independent spirit or soul, a self that exists separate from the body. Now, as the previous discussions of memory, personality, consciousness, and self point out, evolutionary development has provided all living beings with the potential to adapt to their environments by perceiving and interpreting those environments via an amazing array of unique methods of varying degrees of complexity. However, as far as humans can determine, our species is the only one whose brain has evolved to the degree that selfness concepts such as a separate spirit and soul have arisen.

And therein lies the confusion: Self-consciousness occurs because of control requirements, but a separate entity of that selfness does not and cannot emerge. There is nothing in the principles or corollaries of control that makes provision for or necessitates the existence of a separate self. All life forms get along just fine, busily working their control boundaries and adapting as needed, without any need for or evidence of a separate spirit/soul in their lives. Yet, the evolutionary development of physiologically complex brain structure and function makes it possible for such a concept to be thought of, as humans do.

But possibility isn't the same thing as existence. Just because we can think of something doesn't mean it exists, to state the obvious. Human history is a catalog of misplaced confidence in thoughts confused with reality, many humorous but many also disastrous. For example, since approximately a third of the human brain cortex is engaged in vision, it's our most-used sense and therefore requires multiple filtering by cultural, psychological, and physiological processes to understand what is being recorded, which leads to wonderful cognitive ability but also unreliable results. Hallucinations, magic illusions, and other visual distortions by individuals of what they consider to be reality are examples of the ability of an advanced brain to veer wildly across the spectrum of consciousness.

Adding in similar problems with memory as a perishable recording faculty of the brain and human fallibility of perception input, storage, and evaluation, these difficulties strongly confirm that the concept of a separate selfness—a spirit or soul—falls into the category of erroneous thinking as well.

Furthermore, it's important to remember that this cognitive ability for imaginative thought is also paralleled by humanity's near-exclusive possession of the ability to have culture and that culture is transmitted from one individual to another through learning: cognitive processing. Put otherwise, culture is something thought up, which eventually disappears with the individuals who learned it because it has no existence separate from them. (Cultural artifacts temporarily survive, but they are only products, not the concepts themselves.) Likewise, spirit and soul are cultural concepts and therefore have no separate existence outside any individual who has learned such concepts.

Culture is one of the control filters, along with memory, sensory input, and predictive cognitive processing, which determines our interpretation of what we consider reality. If individual members of a social group interpret specific environmental phenomenon to validate their understanding of that phenomenon so they agree there is a separate spirit or soul selfness to objects (including human bodies) involved, then their beliefs are formalized as part of a culture and taught to other members of the group over generations—

until the group's individuals no longer remember or value those beliefs.

Not that the concepts are inconsequential. Just like concepts such as freedom, family, truth, discrimination, and deception, they are likewise consequential. It's just that they all need to be understood what they essentially are: cultural constructs of belief and value, depending solely on their continued existence in a society for their control utility by the individuals in that society. When any concept is no longer useful, it will be adapted or discarded, whether of language, artifact, belief, or behavior—such is the nature of culture.

Reality

All of this discussion brings up consideration of what is out there that living beings perceive and if there materially is anything, as some have questioned. And, yes, this is a matter that humans have debated since they were capable of doing so. But let's not go there. Instead, we should continue our focus on defining according to consistency with the principles of control.

In brief, reality concerns an individual living being's environment of the non-living planetary natural elements and forces, other living individuals, and the influence that all have upon one another as individuals struggle for control to live. In defining reality, then, the key elements of control as an explanation of reality are the necessity and ability of living organisms to perceive and interpret that environmental information according to evolutionary neurological development, memory, and culture.

Perception involves sensory input and interpretation of that input. There are many levels of complexity and modes of sensing and interpretation throughout the entire spectrum of life forms, and some previous discussions have already touched upon those differences. Sensory input is primarily a physiological feature of the brain and body and can be examined with detecting equipment and scientific procedure. Interpretation involves the physiology of the brain and, more difficult for scientific study, the results of the neurological functioning of that physiology, i.e., the cognition of the brain.

For humans, the most influential elements of cognition that affect perception are the thoughts and behavior regarding an individual's culture and variations on the expression of that culture, which results in the unique personality of an individual.

Every person interprets sensory input according to their brain's memory capability of storing a personal history of events and processing thoughts about those past and current events by the brain's predictive coding ability (making predictions about perception at the network and neuron synapse level regarding environmental input and resultant decision-making).

We also interpret sensory input following cultural directives such as concepts, behaviors, and products of those in the forms of language, social associations, beliefs, and artifacts—at a minimum. Some individuals have greater or less genetically determined brain size and information processing capacity than others; furthermore, brain size and processing capacity can be affected by disease or other damage during their lifetime.

Regardless, individuals use whatever they have available to interpret sensory input according to their culture, plus whatever happens to them uniquely as individuals throughout their lives—their memory. This filter of brain, memory, and culture determines what reality is for them.

If a sufficient number of individuals agree on enough observations, then that's accepted as reality by their particular social group and integrated into cultural belief systems which are considered being general knowledge or are formally declared true by cultural institutions (courts of law, governing legislatures, religious organizations, scientific publications, etc.).

It's basically a determination of reality by agreement; and if an individual's interpretation of sensory input is too far from consensus, then he or she is considered not to be in contact with reality. Hence, there's ostracism, confinement, remediation, re-education, and even death as various means for dealing with unacceptable and socially threatening understandings of reality that are at odds with the prevailing culture.

Certainly, the cultural values and institutions of some societies are more accepting and adaptive than others in such situations. Still, the fact remains that reality is both an individually and culturally determined product of statistical agreement within a society (intriguingly not unlike Bayesian statistical predictive coding theory in cognitive science regarding the individual perception process noted above).

Reality for humans, then, is what an individual sensorially perceives through the central nervous system and interprets through cognitive functions of the brain, memory, and a society's culture.

For less complex life forms, there is no culture filter, although learning is still present as part of the cognitive function of the central nervous system; there just isn't any transfer of learning from one generation to the next via culture. Otherwise, the same process of reality perception and interpretation exists for them also, just with less complexity according to the capability of the life forms.

Furthermore, this whole process of reality perception is necessitated according to the principles of control, which requires every living being to be able to sense and respond to its environment by trying to control the environment it can continue living. Without perception and response, control is impossible. Reality perception is therefore logically an element of the principles of control just as memory and consciousness are and, of course, is also integral to their functioning.

Self and Individuality: Community Considerations

Let's take a closer look at that question of what we are: an *us* or an *I*. Because we humans are aware or conscious of a sense of individuality, as imprecise as it may be, this idea of consciousness of self needs to be examined further since it's supposed to be a distinctive trait of humanity. Of course, we have little idea if other animals have it too since we can't communicate cognitively. Moreover, if there are so many individuals and different species of animal and plant life, such as fungi, mites, gut, and skin bacteria, living on us and in us (and by extension, on and in other animals and plants) and they all affect our behavior and also have epigenetic effects, shouldn't they be considered in the question of what constitutes self and individuality if we are to evaluate such concepts?

How much and exactly what the effect is, we don't completely know, but scientific research into the parameters of such community assemblage and control in animals and plants would have to be included in answering the question. It's an intriguing thought that all of our body cells which are communicating with each other via electrical and chemical neurological pathways are also communicating with millions or billions of other individuals of many bacterial, plant, and animal species housed in our body home, all struggling to assert control boundaries as best they can for their individual existence—the result of which we humans conceive of as a level of health.

In a way then, the question of self might have to take into consideration the social effect of our bodies as homes for all these other individuals, much like an ant colony, for example, really comprises the totality of the effect of all the other members of the society enacting their roles. The individual's existence in the colony can't be fully considered separately from the rest of the group.

Are we humans as individuals and social animals any different? Is any animal or plant? After all, in the broadest ecological sense, nothing alive exists separate from other living things but is truly only completely defined by its relationship to other living beings.

But wait. In considering death as the ultimate dividing line between self and body, maybe every one of us communal beings in a single body doesn't die. Instead, some of us (especially bacteria) could continue living for a while and eventually become part of another communal being, such as being ingested by scavenging invertebrates or carnivores. As an example, isn't that how contamination and disease progression works, wherein the disease-carrying organism is transferred to a new home and asserts control there according to its DNA genetic memory directives? So, a part of us humans could continue, perhaps for a long time, in the form of some of our former microscopic community fellows just by living in a new residence. Yes, but there would not be a memory or consciousness of that old communal being since these individuals aren't capable of cognitive memory; that element is an organizing function of the more complex communal housing organism, such as a human being.

Perhaps individuals truly exist only in simple forms such as one-celled beings which are too small to harbor other life forms. Everything else alive may be a communal being that acts in coordinated control contention so it just appears to be a distinct "individual," the definition of which depends on the degree to which observers can distinguish the kind of life forms that make up the total composition—a little adjunct to the process of taxonomy. Microscopes are an aid here (think of the discovery of microbes), as well as careful definitions.

Emotions and Self

Still, it's hard for many to let go of this (controlling) idea of a permanent independent self, of not wanting to lose what impression sentient beings, especially humans, may have of themselves. After all, in some form, all living beings have a sense of self and hence the motivation to keep living; but only those that are capable of reflection perhaps have this problem thinking of themselves as a unique unit and not wanting to lose it. I know that as a small child, the idea of time going on forever unending, and "me" as something that did not physically exist anymore was a darkly disturbing concept I could not grasp.

And perhaps that's entirely the wrong way to think about it: that I lost something, my self. Maybe the problem is just a matter of how we think of ourselves and, importantly, that we are emotionally attached to the whole.

From an individual's point of view, our senses perceive us all as separate and unique, whereas we really comprise a whole interactive relationship of billions of different living units and species—and so does every other living organism.

So, our emotional attachment is actually to this roiling communal environment bounded by skin, hair, and nails. Is that what we want for our emotional attachment? In substance, it would seem so. Then, how does this change our perception of ourselves, or does it?

Until the scientific discovery that we had this communal involvement on such an extensive scale, we rarely considered the situation, other than in the realm of disease. However, since we now have some inkling of the potential significance of the community we each harbor, the question of what "we" really are and our emotional attachment to our body deserves a more penetrating consideration at some point.

Extending the question about emotional attachment further, one might also ask, how far out beyond our physical boundaries does this emotional bonding to self reach? Well, in terms of the control paradigm, it can go out as far as our control boundaries extend, at least the boundaries that cultural and personal history help define our sense of self.

For example, the love of a certain physical environment, such as mountains or a nearby park, could be part of one's sense of self and attachment and be the reason for our desire to protect it. In the same way, a desire to protect one's family or other people would be an extension of our desire to control boundaries we attach to our sense of self and would be reluctant to let go of because we feel it's a part of ourselves. We just hate to give it all up, emotionally speaking.

Yet, while it can be a disturbing thought that an individual returns to the original state of simple atoms upon death, why should it be? After all, before we were born, that's what we were: a bunch of chemical particles, among trillions of others, which just happened to get organized into something living. There was no awareness of any existence before conception, so what is the difficulty with simply returning to that state of being again?

Viewed from another perspective, perhaps the problem with a sense of loss of one's self is rooted primarily in our emotional connection to our ability to be self-aware, self-conscious: a capability which is the evolutionary result of our physiological cognitive development as humans.

Dealing with that sense of loss is a psychological process, as in adapting our behavior to manage our interrelated boundaries of control, and a cultural process, as with utilizing organized concepts such as religion and philosophy for help.

Emotions and Buddhism

As an example of a cultural process for managing that sense of loss and emotion, Buddhist philosophy may be relevant here, particularly in its concept of non-self, anatman: that there is no permanent independently functioning self that exists. According to this view, what we think of as a self, constituting the core of our bodily essence, does not exist; it's a function of our perceptions and desires. The same is true regarding all other living beings and other phenomena we say exist; they exist only in our minds because of our perceptions. The physical elements that embody what we perceive exist independently of us, but the constituted form we organize by perception and thinking does not. So, broadly applying the proper precepts and concepts of Buddhism, we should be able to rid our thinking of the idea of self by accepting the reality of impermanence of ideas and things and therefore have nothing to lose and nothing to be mournful about.

This viewpoint also reflects the principles of the control paradigm: that we, and all beings, consist of basic atomic particles and forces which only come into a living form under certain circumstances, seek to assert control over our environments, and then return to a non-living state of basic atomic particles once again when essential control fails. Since there was no thing (self) there to begin with, there is no thing (self) left at the end of that little adventure in living by control. Out of disorder, order, and then back to disorder.

And the living-by-control episode in between? From the Buddhist perspective, it's a product of our perception and emotional attachment to that perception of what is occurring, giving it memory value and emotional desire which can inevitably lead to unhappiness when it all does not come out as we had hoped. To escape this assured struggle and unhappiness, we need to let go of unobtainable desires and realize the impermanence of ourselves and all that is in the world (the essentials of the Buddha's first three Noble Truths). Then, we will not mourn the loss of that which never existed: a self separate from the perceived body.

Still, neither Buddhism nor the control paradigm concludes that the episode in between birth and death (i.e., life) does not happen for the individual. On the contrary it does, according to whatever cultural frameworks of observation and interaction with which we order it: for example, religion, scientific study, artistic work, music, sport, literary works, or even a life of crime and debauchery.

The problem comes when we emotionally attach ourselves unreasonably to what we want and our efforts for control to achieve it don't work out as hoped. Unhappiness is the reaction to a failed control attempt. It isn't the framework itself that is the problem, rather it's the unmindful and illogical emotional valuation (greed, desire, etc.) that we give to it. Put practically as a solution: If you don't have control, let it go—and escape the suffering.

Put in perspective, the control paradigm is the primary concept which explains the foundation of life and its reasoned corollaries. All cultural control frameworks of behavior and value build upon this concept, including the framework of Buddhism, which I suggest has much in common with the principles of control by virtue of similar assumptions and corollaries and offers a way to deal with a major problem of human existence: emotional attachment.

Further, this does not mean that cognitive frameworks such as Buddhist philosophy, scientific inquiry, religion, or the control paradigm consider emotion irrelevant to living beings; they don't. It's just that emotions need to be recognized as a part of living, understood, and controlled for the benefit of the individual (alone or as a member of a social group) by such means as the Eightfold Path of Buddhism or any other similarly oriented cultural concept.

Emotion Defined

It would seem that emotion needs to be more specifically defined, and its role in the functioning of control examined more closely since decisions are not made solely based on reason, neither in humans nor any other life form. Remember that consciousness is the perception and reaction of the individual to the environments outside and inside the body as it tries to control those environments through bodily chemical and/or neurological activity. This activity involves perception, memory, cognition, and behavior; it also includes emotion since that also permeates all of consciousness.

Most generally, emotion can be thought of as an enhanced conscious reaction to sensory perception of environments outside and inside the body as part of the individual's control activity. This involves interpretation of that perception, according to culture and cognitive processes of reality evaluation and change, and enhancement especially by neurochemical substances (e.g., hormones) which are introduced to heighten or moderate that interpretation and produce physiological responses in both behavior and mental concepts.

Fear, love, happiness, anger, joy, and all other emotions are conceptual and bodily reactions to stimuli. These are filtered and interpreted by an individual's memory of personal events, their cultural values and beliefs, and genetic memory-driven physiological processes that produce models of expectations for what is happening and what can next happen. Emotions thus provide non-logic-dominated motivation for action, which can be rapidly employed and have the potential for more positive control of the situation encountered.

Of course, there's the risk of emotions having a less positive result, which is why we usually have mixed emotions regarding what to do in making some decisions; and our brain, in its evaluation of the situation, tries to balance the mix of emotions and reason for the best outcome. Therein also lies the formula for the human conundrum of our species' history: how to balance the two, reason and emotion. Both have complexly evolved through time and have the potential for that advancement to do great things with the resultant brain power humans now have. We just have to use it wisely in our control activities.

Religion

Mammalian Foundation

An example of this combination of evolutionary brain development and emotion along with cultural beliefs and behavior is the concept of religion. This mixture has been a major force throughout the history of human control behavior. However, as with most cultural beliefs and associated behavior, the potential of this concept as an instrument for positive control has had mixed results. To understand this, we need to look further into the basics of the human organism and its control behavior.

Like all mammals, humans are born dependent upon someone in a parental role to take care of them, and it becomes part of our socialization: an expectation that someone will take care of us when we need it. As babies, our behavior is emotionally instinctive and based on the brain's limbic system control of the endocrine and autonomic nervous systems: for example, instinctive crying to get us what we want. Growing up, then becomes a process of gaining more control of our life through higher functions of learning and less reliance on instinct and emotion.

Also, as we are more successful in asserting control, the less we need or seek the power of the parent or others in a parental caregiving role, such as teachers, the older brother or sister, an older friend, and a mentor at work. Yet, even though we become more in control of our environment through the years and that expectation and desire for parental care diminish and change as we grow older, it never goes away completely. When circumstances are perceived to be far beyond an individual's control, we are still predisposed to seek the control intervention of some power greater than our perceived individual selves.

This can take a variety of forms such as voluntary and governmental institutions like social clubs, police, social services, political parties, or local gangs. However, in cases that seem tremendously overwhelming, especially when emotions are involved, as they usually are, humans often seek some kind of super-power to take over control of the situation for them, just as their parental-role persons once did. A hurricane bearing down, a criminal threatening with a gun, a fatal automobile crash out of nowhere, a surprising disease that crashes one's world of happiness—all are situations that stretch beyond the normal areas of individual control and are circumstances that can bring a person to desperately seek help from a power source such as a supernatural being or god.

The Supernatural

This being or god can be one that's familiar to the person through social institutions such as a church, a loosely organized belief system of rituals, or one that's perceived to arise spontaneously at the moment of crisis by surrendering responsibility for action to that supernatural being—or a combination of these. Regardless of how the individual comes to seek help, the stimulus for and asking of aid is based on learned patterns of mammalian behavior and neurological brain processing. These filter and direct religious cultural concepts to become the means for an individual's attempts to control whatever part of the environment it needs to control, if possible: whether a tornado of the non-living environment or a marriage disaster of the living environment.

Like any other human behavior which is directed to solve control conflict, the action of seeking help has some degree of hormonal guided emotion involved which provides not only added stimulus for action but also an effect upon neurological predictive processing for determining the reality of perceptions. When encountering extremely difficult situations, our perception of the circumstances involved is colored especially by previously held cultural concepts and beliefs, including a personal history of emotions, which in turn was previously colored by cultural concepts and beliefs and so on in a feedback loop. Besides this influence on individual perception, there's the contribution made by the perceptions of any group members involved in the situation.

As analyzed earlier, the process of reality determination is based on agreement among participants as to what is correct and what isn't, whether a certain tree or mountain has a spirit with supernatural powers or a certain person has such powers because of recorded or remembered stories of that person's teachings or actions, for example. If enough of the relevant people agree that it's real, then it is for them but isn't for others not participating in the agreement process.

This is how all cultural concepts and beliefs work: for those sharing the culture, it's reality, at least until it's no longer practically functional for the individuals in the social group. At that point, the unneeded cultural elements are abandoned, adapted, or replaced by better models for behavior. The only function of culture is for establishing control boundaries, and that includes all elements of culture;

it's humanity's most powerful tool in the struggle with non-living and living environments.

Accordingly, how does religion fit in with this struggle? Quite simply. Since it has roots in our mammalian evolution and predisposes individuals to bond with others, including to seek help in solving individual control conflict and has enhanced emotional motivational value in those efforts, this combination gives the cultural concept of supernatural power an important role in most societies.

This isn't the same as saying that the concept has a basis in any reality outside that society and its members, since it does not, as discussed earlier; it only has a potential role in the attempts of individuals to control their environments—just like all other cultural concepts do to varying degrees of effectiveness.

The primary difference with the concept of religion compared to those of democracy, freedom, equality, and authoritarianism, for example, is that religion is the only major concept primarily based on a belief in the power of supernaturalism. In contrast, all other concepts are based on human capabilities alone and behavior accepted by the consensual agreement of a society's members. This inherent difference is a source of potential conflict within all societies since it divides their members into those who believe and those who don't (unless all members agree to the same religious concept, of course).

Cultural Foundation

The focus here isn't to examine the relative merits of religion and other concepts but to consider its role regarding control for the individual. This has been done previously concerning mammalian origins for motivation and supernatural intervention: seeking support for physical and, especially, emotional needs.

Humans, of all mammals, have the additional advantage of a capacity for complex cultural support in aiding individual control efforts, primary of which is the aid of social groups based on specific concepts of family, kinship, economics, political power and government, and voluntary organizations. These groupings can help the individual utilize resources of the non-living environment and the living environment, including other human members of the living environment. Given this basis, religion is yet another cultural concept that is available to aid an individual in generating needed control.

Another important element of religion as a cultural concept, besides the emotional enhancement and belief in supernatural power, is the more encompassing control vista that it usually assumes. It tries to aid the individual in controlling many more elements of the environments than other concepts do, either through providing an ideological understanding of the situations or by offering guidelines for behavior to aid in the control situation.

Other concepts likewise try to aid through understanding and guidelines for action but are more limited in scope, seeking to aid in controlling specific economic, environmental, familial, medical, or other areas of potential conflict. Many also use emotion as an important part of membership involvement. Of course, there are many other concepts and social organizations based on those ideas which likewise attempt to broadly explain and aid in individual control attempts, but the combination of emotion, supernaturalism, and broad worldview make religion a unique cultural concept regarding control.

Now, two observations about uniqueness are to be made. First, we must keep in mind that religion as a cultural concept is an invention of human origin just like every other cultural concept and can change or disappear just like every other concept. For example, if the assumption of supernatural beings or powers was no longer deemed acceptable by consensus of membership—either by the totality of humanity or a particular group—then that element of religion would disappear and religion would then rely solely on its emotional aspect and broad worldview as an appeal to adherents of the concept.

Some might then question whether religion as a unique concept could continue without the supernatural element, with its origin from mammalian attachment and seeking such care in times of need. Without the solution for that need being transferred to supernatural help, could religion continue to exist? It's a legitimate question to ask. Of course, it's conceivable that emotional support between members of religion just based on mammalian care conditioning and the worldview of trying to help with a broad range of control issues might be sufficient to make it an effective control framework. But then, why call it religion?

There already exist such organizations based on mammalian care conditioning, such as crisis interventions, environmental and animal protection, drug and alcohol addiction centers, legal aid, food distribution, health care, and on and on. Some are quite narrow in scope, while others are extensive in function and geographical distribution. Even a national government itself and international institutions can be considered being concepts originating from mammalian social instincts and focused on promoting individual control over a broad range of concerns. So, the key element, the uniqueness of religion, does rely on supernaturalism.

Second to keep in mind is that, regardless of the origins and beliefs of a particular cultural concept, all cultural concepts must be evaluated on the same criteria: Does the concept genuinely help the individual's control endeavors, either directly or through a group effort? Essentially, this is the stuff all social sciences deal with and try to explain; and again, examples abound to be investigated by those interested. Here, we will discuss only the general variables that religion, like other concepts, encounters in attempting to perform this control function and the associated consequences.

Religion does benefit individuals directly and through group action. As a concept, for whatever combination of reasons, it gives some individuals guidance, emotional fortitude, and social benefit, just as any cultural concept can for members of society. Yet, by being a concept which only certain individuals can adhere to because of the requirements of particular beliefs, it's an exclusionary concept, just like any other concept which has specific belief requirements. This therefore also makes religion susceptible to many difficulties in carrying out its role as a cultural framework for aiding an individual in control behavior. Remember, culture exists only for that purpose, and therefore religion also exists only for control.

Membership and Exclusion

A major problem in control effectiveness, then, is the reaction of individuals and social groups who don't take part in a specific religious concept. If in the minority, religious members can be forced to convert, be persecuted, dispersed, or killed. If in the majority, religious members can force non-believers to convert, be persecuted, dispersed, or killed. Plus, any combination of those behaviors is possible in either situation. Moreover, this can and has happened throughout human history in other social situations where believers in other exclusionary concepts have been in a minority or majority membership, such as political, economic, sexual orientation, ethnic, racial, and various other groupings.

It would seem that this feature would be very much counter to a positive control purpose of religion and any other concept. Yet, not all religions, nor other cultural concepts, only have dire consequences, and examples abound throughout history which show that they can be useful frameworks for individual control endeavors.

However, one could also ask, is the situation the same with concepts that don't have an exclusionary basis for membership? Logically, it would appear not to be. For example, freedom would seem to be a concept that would be open to all. Who could argue against freedom? Or equality, the rule of law, justice, peace, choice, and the common good? And in fact, there must exist a minimal level of accepted non-exclusive concepts by the members of a society for it to function, anyway. It's even conceivable that these concepts could be agreed to by all humans on Earth and be used for the benefit of all individuals in the struggle with non-living and living environments. All for one, and one for all.

Unfortunately, there are other cultural concepts that undermine this potential, such as slavery, monarchy, dictatorship, genocide, monopoly, and revolution, that have exclusionary worldviews and are susceptible to using emotion for an enhanced effect. The key here, of course, is that all these positive and negative (depending on a society's values) cultural concepts exist only for control.

But individuals have different control boundaries, different perceived requirements of what they need to control, and to what degree, so they all seek to employ whatever cultural concepts are most useful in establishing those boundaries. Hence, the problem for idealistic values is to strike a balance between the reality of cultural capability and the requirements of control.

Additionally, although all religions have inclusive concepts, for example, peace, tolerance, and kindness to promote a unity of membership, they also utilize exclusionary ones as informational tools to attract potential members. They must have something to offer which others don't have; otherwise, why choose them over the competition? Of course, this leads to proselytizing by many religious faiths to gain members and increase control, just as with many non-religious organizations, and can lead to positive benefits for the individual or be an annoyance at the least and turn to persecution at the worst.

Given this reality of membership, humans have a choice not only among religions but also among branches of those religions or adaptations of them. Furthermore, with the special attraction that religion has for emotional behavior, this competition for membership can become quite extreme, just as it can be with many other social groupings that rely on emotion and various exclusionary cultural concepts other than supernaturalism. Thus, religion, like every other cultural concept, suffers from the many control problems that culture brings to the fore but also enjoys the benefits control solutions provide.

IV

DIALECTIC DIALOGUES: Looking Beyond

———————— ~——— ————————

Purpose

In human discourse and observation of behavior, the question of purpose in life often comes up, particularly in the realms of science, religion, and philosophy, in both professional and simple commonplace discussions about daily life. Our emotions fluctuate between highs and lows as we evaluate what is happening to us, and at some point, one's mind brings up the question of why you're doing what you're doing. What's the purpose of, well, everything?

On the most basic level, every living being has physical survival as a purpose for living because to stay alive one has to have a successful domain of control and boundaries; that's the fundamental principle of control. So, survival is the purpose, and that means control. OK. Anything beyond that? Control beyond basic survival and all the factors involved would seem to be the next layer of purpose; and depending on the capabilities of the living organism, those can be simple or complex.

In simple living forms, just existing is a sufficient purpose for living, an appendage to mouth existence, so to say. After that, complexity increases tremendously for all living forms. Looking at the end of the range, human culture and bodily neurological capabilities can make the purpose of a control event very complex, since cultural values associated with behavior are part of control in a self-reinforcing feedback loop. Whatever values an individual or group holds provide purpose and neurological stimulation for action—which reinforces those values and sense of purpose, resulting in stimulation for more action, and the reinforcement cycle repeats.

Examined more closely, though, the reward for value fulfillment (and purpose) is also a matter of chemistry, of having hormone secretions such as dopamine and serotonin which give emotional pleasure and reinforcement for direction and simple bodily functioning. For example, if adventure travel, family relationships, or addictive drug use and the stimulation from those is sufficiently rewarding in emotional pleasure, then that would provide a purpose for continuing the act of living, to put it simply.

It could be one's reason for living, at least as part of a set of pleasurable actions; and as discussed earlier, emotion is a key element in motivation and decision-making in general, and establishing purpose as well. This makes sense logically, though it does seem sort of ignoble, in some ways, to have hormone reactions as a reason for living. Yet, those are what help drive the basic motivation of all living beings for physical survival by aiding in the control of an individual's many bodily functions such as food processing, breathing, and flight response.

Hormones also directly affect the many cognitively and emotionally based choices made which will aid in survival—decisions which contribute to the motivation for living, particularly for more complex beings, including humans (e.g., when to join a particular social group, obeying or breaking laws, and seeking comfort from another individual). Some are simpler selection processes than others, but no less important.

On the individual human level of self-awareness, we can then conclude that beyond basic survival, the search to find purpose and continue living also depends upon finding hormone-producing activities that are emotionally rewarding and make the individual want to continue them. Furthermore, cultural values and personal history will provide guidance for finding and integrating those activities, and one needs to examine those to aid in making choices and help reach the goal of having a good life.

This sounds reasonable and practical, and of course, it is; but obviously, it can be a challenge to carry out, as the discussion on emotions and Buddhism underscores. Yet, understanding and appreciating the encompassment of the control paradigm—at least as to who, what, and why specific factors are involved in a control event—can be an aid in making those choices of purpose for one's life.

Is Control Always Self-serving?

Yes, by definition, it is. As discussed above and elsewhere in this treatise, the purpose of control behavior is to enable the individual to continue living, meeting physiological, cognitive, and emotional needs present while the individual is alive. This seems elementary, with minor disagreement possible regarding all life forms. However, in more closely detailing the observed behavior of humans and various mammals, additional considerations arise to clarify this matter.

For example, we have words that seem to embody motivation that isn't self-serving, such as love, altruism, self-sacrifice, compassion, friendship, generosity, charity, and so forth. At the other end of the spectrum, antonyms for these would be self-serving; so is it one or the other, or in between?

As hinted, there is a spectrum of degrees of self-serving between the extremes, but there's always some element of selfishness that must be involved in the activity since that's the primary underlying motivation for any control scenario: manipulation of the environment for individual benefit. The proportions of selfishness vary according to specific cultural concepts which the individual believes and follows, the personalities of other individuals in the behavioral transaction, the particular personal bodily interior environmental emotional states the individual perceives at the moment, the relationships with any organized social organizations involved, and any other living environmental association that's contingent upon the control situation.

It isn't a simple process to decide where any particular selfishness lies on the spectrum. Thankfully, as with the vast majority of decisions that constantly must be made every day, habituation relieves the individual of reprocessing the total activity again and again. We decide that the result of a particular control activity is beneficial and then go on with our day and take a similar action without another thought when the situation arises again. That's how the vast majority of decisions are processed, roughly speaking. Yet, the self-centered element regarding more thoughtful decisions needs further elucidation.

It's still fair to ask, "What's in it for me?" when a charitable contribution is sent, or an out-stretched puppy head is patted, or a child is comforted after falling, or a civilian rushes to a flaming car crash and pulls out the unconscious driver. What's in it for the doer? The answer lies in the influencing contingencies on that spectrum of selfishness as mentioned above; each situation has a variation on how the actor benefits. Regardless, an individual must always gain something which enhances their perception of themselves.

This can take place concerning their cognitive awareness of self or their emotional reactions involved in the control event, or a reinforcing combination of the two. The greatest influence is upon the emotional contribution: the various hormonal contributions involved in the physical activity of the control action. Oxytocin makes it just feel good to pet that pup or console that child, and the added adrenaline rush of excitement from a car rescue is something an individual will never forget—all of which also add to the concept of one's self and personality.

So, control always involves self-serving even if primarily for the emotional reward, because control is the contributing motivating factor in the activity, to begin with. An individual must always get a benefit of some kind and to some extent, otherwise, there will be no basic, instinctive stimulus to take any action at all.

Ritual and Habit

Given that when we die there's nothing left except some atoms and molecules to rejoin the rest of such on Earth, there isn't anything we can do about that result. Death is inevitable for every living being. Now, this process of dispersal can be disturbing to us and is likely to also be for any other cognizant beings. Yet, depending on how you look at it, this can be freeing instead of confining because it releases us from the feeling that we are responsible for many things in life that we believe are important for the future. In fact, what we individually do in life rarely has very little if any effect on the future of anything in the grand scheme of time, and affects only a small segment of an individual's immediate present. Encouragingly, that small segment is something where we can genuinely exert an influence.

This brings us to think about the purpose of ritual in life, formal and informal. It would seem that ritual is a way of trying to negate the feeling of inevitability of death and consequent powerlessness to have any effect on the future, to escape a sense of meaninglessness, and to give a sense of power to one's actions and to the beliefs which underlie those actions. Rituals can create the feeling that there's continuity to life and an ability to control it, whether it's prayer, a superstitious motion, formal ceremony, or whatever is done in a format of repeated action in a particular environmental surrounding. Thus, the emotional sense of being helpless in the face of death and dispersal can be neutralized to some extent by transferring, through ritual, our attention to those areas of living where we have more control.

The biological derivation of ritual comes from repeated actions of daily living, such as automatically dressing oneself in a certain sequence and from other daily procedures where some routines have been found to work best and are assigned, consciously or unconsciously, to habituation, which in essence is a control mechanism of the brain to efficiently expedite decision-making regarding daily living routines. Ritual, then, is just an extension of habituation, with an initial conscious action and the added expectation of a reward beyond that behavior.

If there are enough occasions of getting a perceived reward, then the action can become ritualized. The reward does not have to be large or very frequent, just enough to establish an integrated cognitive and emotional reaction that's sufficient to establish the ritual, a sense that it just feels right besides the hope for a reward. Simple examples might be wearing a favorite piece of clothing or accessory because it seemed to go along with having had some good days at work, or not leaving the house for the day without a kiss for the spouse since the day you didn't, he barely escaped a traffic accident. A classic example is in sports or other events of skill where certain activities before activity engagement are sometimes obsessively carried out, such as preparation rituals of hand motions, clothing adjustment, and equipment checking before an explosive release of energy, such as in baseball and tennis.

Thus, action and reward easily turn into ritual since we typically tend to remember the good more than the bad and assign a positive outcome of action as due to the ritual. Interestingly, ritualization is a similar feature in some non-human animal behavior too, generally with those having a more complex evolutionary development such as elephants and primates.

It's easy to see how this informal behavior can be extended to cultural concepts of ancestor veneration, magic, and religion where some power beyond the individual is sought through ritualized action or speech. Thinking more widely, this behavior can also include more formalized group behavior and social institutionalization, as in religious places of worship, liturgy, and group singing or speech. These all fit well with the concept of control since such activity provides the individual with a perceived means of control over personal boundaries which can include just themselves or be extended into social boundaries involving direct control over many people, as might be done by politicians, religious persons, warriors, and so on. Regardless of whatever level at which the ritualization is carried out, its motivation and purpose are for control; any other cultural purposes are an extension of that principal objective.

Sexual and Asexual Reproduction: Considerations

The role that sexual reproduction plays in directing more complex life forms toward associative or communal behavior is interesting. Examining the evolutionary tree, we find that the interaction of sex and associated mammalian characteristics of attachment through parenting is necessary to help explain human social behavior. The possession of sexual attributes logically means the necessity of interaction with other like beings and more opportunity for genetic variation in evolution. When combined with our capacity for culture, this can lead to both beneficial and destructive results. Overall, historically, it would seem that keeping a balance between sex and social attachment is something else that humans as a species have struggled with throughout their evolution.

Interestingly, in the primate family tree, bonobo apes may have avoided this development, in contrast to all the other apes and humans—and probably most other mammals and less complex animals as well. They use sex to positively cement beneficial relationships instead of destructive ones by being promiscuous (no animosity from attachments), equally sharing brief sex acts (as a greeting and reassurance), and female group leadership. This all leads one to wonder if asexually reproducing life forms, by not needing to interact sexually with others for reproduction, avoid the problems resulting from the need for association, or can it just be avoided by some means such as the bonobo have?

This, interestingly, brings up the possibility of life forms elsewhere in the universe perhaps being predominantly asexual and having evolved into more advanced forms—an evolutionary family tree but without sexual differentiation following a male or female line of descent. What would the contrasts be, if it were even possible for them to develop into complex beings similar to humans?

Logically, it would seem that nothing in the control paradigm would preclude such a development, but there would be a high likelihood of extreme slowness of evolutionary change due to relying solely on epigenetic or random genetic variation during cell subdivision and ensuing replication, instead of the additional variation potential offered by the fusion of different gametes in sexual reproduction. It may be that on Earth, that's simply the way it turned out for us, and elsewhere it may happen to be different.

History, Humans, and Others: Is There Hope?

In thinking more in the long term of time, about climate change and consequences, the general direction of human history, and evaluating the positive and negative of it all, one needs to take some perspective. Some analysts posit that this is currently the most peaceful and favorable time in history, using statistical data to validate their assertion. That conclusion is rather simplified, however, since ignoring those who could not benefit from any peace or wealth still number in the hundreds of millions, or billions. We must evaluate this issue based on the individual, not solely on a statistical number; perception is of key importance since peace and prosperity are relative. As we all know in our daily lives, it's little comfort to know that others are doing just great while we're at the bottom of an emotional and financial pit with the world falling in on top of us. An individual needs a control framework that offers hope of change from the undesirable to the desirable. Given human history, is that possible?

In many respects, it depends on where you live. In thinking about different locations around the world, it's quite difficult to think of any place that has not had a major human conflict that comes at least every generation or two. Going back over European history, for example, there has always been conflict, even in prehistoric times, as archaeology shows. The same could be said for the Middle East, Africa, Far East, and even more isolated Australia and the Americas, though the latter two might be regions where the time between band and tribal conflicts was perhaps greater since they did have vast uninhabited continents to move into.

At first glance, it further could be suggested that the only continents in the world where the displacement of existing human groups by others did not occur were Australia and the Americas since the aboriginal peoples migrating from the Asian continent were the first to get there and occupy the lands.

However, later on in both situations, Europeans also thought they were moving to uninhabited continents, but of course, they weren't. It was just a mindset, a cultural construct of control defining the meaning of humanity which was used to justify the near extermination of the native peoples. In that way, the population displacement in Australia and the Americas was no different than anywhere else in the world.

Therefore, as a closer analysis of the examples above show, a lack of conflict only holds for first-comers; every group thereafter has to manage their boundaries of control regarding every following group. Once someone is there, everyone else is a newcomer in a never-ending series of migrating waves. This would hold for Asia, Oceania, and Europe as the various human groups moved out of Africa into those areas and beyond into Australia and the Americas. Everyone after the aboriginal firsts were newcomer migrants, and both had to deal with control boundaries, inevitably leading to various degrees of conflict.

Taking recent history, for example, many US citizens have a distorted viewpoint on serious homeland conflict since compared to the nations from which people immigrated to reach North America and the frequent wars they escaped, it appears that the US has had little warfare on its soil. To be sure, this is a matter of perspective. The nation superficially has not had to live with invaders from outside because of its semi-isolating oceans, in contrast to most other countries.

But of course, Native Americans throughout the continent truly did face invasion from overseas by European nations seeking resource exploitation and land to claim; it just was not forthrightly recorded as such by the invaders and their descendants. Moreover, later battles with and displacement of those aboriginal natives, revolutionary separation from England, the establishment of slavery, controversial acquisition of territory from Mexico, and the Civil War between the states were also very substantial control conflicts that took place on US soil. And their consequences continue in many serious ways to give individuals personal pain and to disrupt positive efforts made by society to change those results: for example, the racial and ethnic discrimination that still permeates the nation. The sum and substance are that human conflict between groups, which in turn causes individual suffering, is the norm throughout all of human history, regardless of location on the planet. So, what is the broader significance of all this?

Is Conflict Inevitable?

Control, as a requirement of living, means conflict between individual beings and groups of beings, whether animal, plant, or bacteria. How that conflict occurs and is handled by each species' control processes is the essence of the species. The crux of the matter is that their "history" of control conflict is recorded through genetic adaptations and/or cognitive learning by an individual being, which is imprinted as memory in a simple nerve network (such as a roundworm) or a brain (such as a hungry lion) or, additionally for humans, as a cultural artifact (clay tablet, paper, digital storage, etc.). So then, is there no way to escape this conflict situation for living beings?

To begin with, one primary conflict occurs because we all require energy sources (food) to continue living, and that would seem to mean we all need to eat one another—put in the bluntest terms. However, is that true? Well, plants are largely exempted since the vast majority of their species use sunlight and essential chemical elements, primarily minerals, to make their food and cell structure. A few species do get essential elements from the decay of other plants and animals, but that isn't a process of eating other living beings.

In contrast, animals need to ingest existing organic compounds from other life forms by killing or scavenging them, since animals can't use sunlight directly to make carbohydrates and other essential nutrients as plants do. Therefore, plants are the ultimate source of food for all living beings, either directly by consuming plants or indirectly by the consumption of animals that eat plants (except for some bacteria which use inorganic compounds instead of sunlight or organic compounds). So it seems that plants, unlike animals, can escape some conflict necessitated by control because they can directly use non-living energy sources.

However, there are other primary opportunities for conflict that exist. Part of environmental control is also having an optimum place to live, including space for essential activities, resources to utilize, and protection from predators and weather. Although plants do have one less source of control conflict (food procurement) than animals, they still have these other wellsprings for conflict crucial to individual survival, just as all animals and bacteria do. Even Earth's extremophile species of life found in deep-sea vents, methane, glacial ice, and deep rock habitats still have these environmental sources of conflict that all life forms have.

Thinking further afield, if more advanced forms of life based on direct mineral and chemical energy sources rather than consumption of other life forms could develop somewhere else in the universe, they would still have these same other environmental bases of conflict this planet's animals, plants, and bacteria have, such as resource procurement and protection. Though it would make an interesting science fiction plot, it does not seem that there would be any way for such life forms to completely escape conflict—just like the rest of us on Earth. We all seem to be stuck with that "curse" of being alive, to live in perpetual conflict or potential for conflict. So, how is a living individual supposed to deal with this situation?

Resolving the Conflict Conundrum

The solution would appear to be that individual beings should try to avoid conflict whenever possible. It sounds simple enough, but the stickler is "whenever possible": i.e.

how to evaluate situations and take effective action. One solution is that evolutionary changes and development have provided the main processes for doing this through the directives of our genetic DNA. In other words, selection for running fast, having big teeth, and feathers to allow flying are genetic adaptations for avoiding conflict (and creating it).

Another solution is learning to make individual choices and act on those decisions, which is a function of neurological structure complexity, processing power, and brain size. Being able to initiate group social behavior, develop communication ability, and remember migration routes, for example, all rely on brain physiological and cognitive development and can aid in conflict avoidance.

A final one is the ability to transmit conflict information to other individuals, and groups of individuals, through time via culture. The development of tool use, recording of communications, symbolic artifacts, social groupings, belief systems, etc. are relevant examples of culture regarding the potential for conflict prevention. These are three ways that living beings can avoid conflict, depending on the species and their evolutionary development.

In analyzing these means for avoiding control conflict, we can conclude that plants primarily use genetic information to grow and react to the environment, such as from predation threat. Beyond that, it's hard to say they can make actual choices between alternatives rather than simple stimulus/response behavior unless the concept of "neurological" is defined much more loosely. Animals use genetic and brain choice processes and some also utilize culture. Humans and some primates use culture, though scientific investigation may someday apply the concept more precisely to some cetaceans, birds, and other mammals. But simply having these processes does not guarantee conflict avoidance, just potential avoidance. Is there some leap of concept that can provide a guide?

Maybe such a possible guide could be ethics: a conscious decision to limit conflict based on such values as the greatest common good, efficiency, kindness, sincerity, compassion, honesty, and respect for others. Clearly, this would apply only to life forms capable of culture since those without the required brain complexity don't seem to have the capacity for the application of mental concepts such as this. Still, despite this capacity and given all the evidence of history and science for the potential to avoid conflict, it appears to be impossible for cultural beings to avoid all conflict, at least for humans.

There's another equally essential part of the control paradigm that must be kept in mind. Not only must life forms protect themselves from elements of the non-living and living environments by avoiding conflict, but they also must of necessity seek out the needed resources for their survival by initiating situations that inevitably result in conflict, especially concerning food. This puts humans, and all our fellow Earth passengers, in a tough spot. What to do?

It seems that we continue to do what we have already been doing for all of time: adaptation by genetic evolution, learning, and culture. The key is to control one's exposure to conflict, sometimes avoiding and sometimes seeking since it's a basic necessity for the existence of life. To do this, all life forms use genetic selection, many use decision-making through learning, and mainly humans use culture besides.

Given this situation, cultural beings at least have a greater chance of directing the degree of conflict their control decisions involve, which would have the potential of also reducing some existing conflict in interaction with and between other living beings. And the potential for doing better is good, given history and science records which document attempts to rise to the challenge. Yet, overall, we're basically stuck where we are and have been for hundreds of thousands of years: We know how to reduce and balance conflict, but don't seem to do well at it. How can we do better? Just keep trying as we have, or is there something that has been missed?

More Ruminations and the Future

Although philosophy and other cognitive frameworks such as religion, democracy, science, kinship, and authoritarianism, offer formats for control, the historical record of their erratic positive implementation still forces the question to be asked: Can humans ever escape the existential demands of control that always lead to conflict and being our own worst enemy? One possibility is that perhaps we might achieve it via genetic manipulation of human physiology. Because most animal actions, including human, are dominated by their genetic code directives rather than cognitive development and culture, then human genetic code might be manipulated to eliminate those impulses which lead us into conflict with each other and other life forms—sort of like implanting a "non-aggression gene."

Now, maybe some species on Earth have already done that through natural selection, and we don't know it or at least have not thought about it from that particular perspective (bonobos?). Along these lines of thinking one could contemplate the nature of other advanced cognitive beings that might inhabit other planets, what they would be like, and that if they were like humans, then they too would be erratically exploitative and best be avoided, unless they had done something genetically to alter that inevitability.

However, if control is intrinsically necessary and conflict at control borders is inevitable, then even if a cognitively advanced species was genetically changed to avoid conflict, it would surely cease to exist because the rest of its living environment would take advantage and eliminate it. The necessity of control and borders would seem to make genetic manipulation by such advanced beings counterproductive.

Yet, perhaps the issue isn't to eliminate conflict but to reduce it so coexistence is more possible for them, and they will not destroy themselves and other species at the same time. It would appear that genetic change could provide the potential for that. After all, it was genetic selection that made our cognitive ability possible in the first place. Thus genetic manipulation should be able to eliminate or change some genetic features which would enable humans to become less destructive.

As suggested, perhaps some other species have already done that elsewhere in the universe. There are lots of scenarios to examine here: for example, other life forms helping us genetically change, conflict of the changed versus the non-changed in the universe as the next step in evolution, and who controls whom.

So a question would be, does deliberate genetic change to establish automatic cooperation and remove conflict in a species make the concept of control not fit, that such manipulation takes us beyond control? No, since cooperation itself is still control, and genetic change whether by direct engineering or natural selection is just one way of setting the boundaries of control for a life form. Still, if such attempts at manipulation were to be made, what kinds of genetic changes are necessary to make humans less destructive, and how is that to be decided? This is a thorny but entirely reasonable question to investigate.

Along this line of inquiry but taking a different method of reducing conflict, one might suggest that the way to protect our species from itself is through social control, by developing a specific cultural assemblage of rules, laws, and educational systems that set out boundaries for individuals which allow personal control to a degree but includes social boundaries beyond which individuals must not go so the collective whole of the humanity isn't jeopardized. Sounds like a good idea: protect other living beings, the environment, our species, and individuals all at the same time.

Except, isn't social control what the entirety of human history has been about? And how has that turned out so far? Maybe there's a system that would work, but none to date has been completely successful. Possibly in conjunction with genetic modification? Perhaps built on something like elements of Buddhist philosophy and other similar conceptual frameworks? Conceivably, there's even some other simple concept, such as control seems to be, that could work as an organizing principle? However, maybe it's impossible, after all, because of the fundamental necessity of control for living beings.

Another prospect would be to eliminate at least one source of uncertainty in control: emotions. Our complex brain requires predictive processing to reduce as much uncertainty as possible in sensing the external and internal environments. However, the body's internal environment itself, primarily through emotions, introduces a factor of uncertainty in the form of neurochemicals and response to them which is difficult to eliminate. They are integrated into our entire physiology, and particularly the primitive brain stem regulating basic autonomic bodily functions. How could they be eliminated? Or should they be? Hormones not only increase uncertainty because they temporarily subdue or amplify other elements of the general cognitive process but also have the utilitarian evolutionary control function of aiding the survival of the individual through sexual attraction, flight or fight response, stress reduction, immune system response, sleep cycle, etc. Do we want to get rid of those? Perhaps control of emotions rather than elimination would be better.

This could be done through physiological adjustment of the body's neurochemicals directly through drugs that modify the secretion of chemicals, such as in medication, recreation, and addiction. So, instead of relying on genetic and cultural modification to eliminate control conflict, maybe improved modification of emotional behavior through drugs would be a viable alternative. Or, perhaps overall modification through a combination of genetics, drugs, and culture would work (a la *Brave New World*)? It seems like a huge task which upon closer examination might show that we have already been pursuing that course but not in a planet-wide, multi-species, organized, or intentional fashion—at least one in a positive direction.

An Ecological Viewpoint

But to persist, is there actually no way to confront reality and make our behavior implementing control positive and beneficial? This truly is the most important question for mankind, and any other organisms on this planet and any other planet, since this is a reality that comes with the existence of any organism, regardless of habitat location.

It would seem that the answer in its simplest form would be to view ourselves as a part of the greater whole of the planet, and universe, and to understand our role in its ecology and our particular capabilities of enhancing that ecology for the benefit of all. We need to embrace the intrinsic principles of control and employ a scientific approach to understanding the ecology of our home and how we fit into that ecology, and then establish cultural value guidance for action. That gives us the control that we as individuals and society need as part of our legacy of being living organisms.

But, not to be overly redundant, isn't that what we have already tried to do, at least some of us, to some extent? Have we not tried to live in harmony with nature and each other? Yet, look at all the problems we still have. What could be any different from the situation we are now in? Maybe humans have already been through this cycle many times and our history has already answered this: that it does not work, and we have failed. Perhaps the demands of control for an organism's existence are so strong that our ability to face those demands and adapt our behavior to ecologically-based concepts is impossible, and impossible for any organism anywhere in the universe—which leaves us with visions of discovering enlightened beings, hopefully, to be found somewhere out there in space, enshrouded in the realm of fairy tales.

Possibly we will never rid ourselves of this conundrum of living which suggests such great potential yet demands a struggle against ourselves individually and culturally. After all, Spock and his fellow Vulcans of *Star Trek* could never rid themselves of emotions; they only learned to control the expression of those emotions as best they could and merely appeared to be emotion-free in their behavior. So, the implication would seem to be that humans, or any sentient life form on this or any other planet, can only hope to discover the best techniques of controlling detrimental behavior through an exploration of cultural concepts and values that will counter negative behavior.

Vigorous implementation of the scientific concept of ecology would go a long way toward achieving this goal since it stresses the needed balance of the physical environment and its inhabitants and could guide human individuals and society for achieving that balance. Moreover, the utilization of ethics and rules of law to direct the application of ecological concepts on a world-wide scale is also needed, given the serious results of humanity neglecting its role in the ecology of the planet's functional system. The key to it all is a factual understanding and discerning appreciation of how our world operates, what the precise state of all life forms currently is, and planning a strategy for what needs to be done, which is additionally an educational and communication task.

Since human society is complacent regarding change and seems reluctant to do what is needed despite warnings by informed authorities, it usually takes a perceived crisis to motivate us to heed those warnings and take immediate and sustained action to save ourselves. It's probably the case that a crisis has always been the needed motivator for humans (and any other living being) to aggressively reconfigure their control boundaries, individually and especially as a group. Climate change and pandemics are historically and currently the most potent motivators and will force us to confront necessary world-wide control issues whether we want to or not. So, why not put that human potential accumulated over millions of years to work and directly address the control adaptations that are needed?

Humans must understand that planet Earth will survive anything we do; that isn't the issue. Great species extinctions and massive geographical changes have come and gone over billions of years, and we humans are only an instantaneous flash in the geological time frame of the planet. No, it's the survival of our species' life character on Earth, and the total living and physical environment as we know it—the world we like *as individuals*—that's at stake. If we can't keep what we now have or could have, because of a poor job of control, then we must make vast adaptations; and the transitional consequences of those adaptations will not be pleasant for every individual human, to say nothing of our planetary companion species.

Artificial Life

In any discussion about the future, or the present for that matter, the existence of artificial life (AL) relating to control must be confronted. The idea of artificial life or robotic mechanical structures has been with us from the ancient Greeks to the automaton curiosities of the 1700s and the practical machinery of the Industrial Revolution's weaving looms and prototypical calculating engines, and on to the science fiction media of recent times.

Today, artificial intelligence (AI) is playing or is anticipated to play a role in nearly every aspect of human life. The concern here is how, or if, AL and AI fit with the concept of control, and subsequently, the existence of all life on this planet.

Since the essence of life is the confrontation of the challenges brought by attempts to maximally control elements of the non-living and living environments to continue a life form's existence, it's a matter of determining if AL and AI fit that model as an individual life form. Certainly, perception and interpretation of environmental stimuli is a task of AI, and the processing of that information according to predictive coding for memory storage and further decision-making algorithms is similar in general terms to that involved in natural life forms, humans in particular. So, that part of control appears to fit well.

The other major requirement of control is that the individual must maintain a minimum degree of adaptation to environments if the life form is to continue living, or exist. And that seems to be a key question: is life the same as existence, are they different somehow, or is it merely a matter of word choice? Before an organic life form is alive, it consists of elemental particles that combine in a growing and potentially reproducing organization of control; when it dies, the control and organization end, and the materials of the individual return to elemental particles. That sequence is what we call life for natural living beings. Is it the same for robotic mechanical structures? Can it be called artificial life, or is it all just life?

In considering the process of life formation and death described above for organic life forms, it's hard to see how it might not apply equally to an artificial structure. Broadly put, that's how all mechanical structures begin, operate, and then cease to, well, exist, (since the word "live" seems awkward and out of place). Although artificial structures need outside human help to get started, then maybe that's an important difference. However, no human pops out of nowhere without the help of parents—outside help—so is it genuinely an essential difference? Perhaps, it's just a matter of customary word choice and not a difference of concepts: "life" and "exist" are just basically interchangeable, depending on literary usage.

Yes, maybe, but somehow it seems there's something deeper here, something regarding cultural values and beliefs. Certainly, it's an issue about values, but further scrutiny is still warranted, especially given the human record of blindly following the unexamined path and the potential significance of this technological development for the present and future. Particularly, what is so unsettling about the use of the word "life," beyond the cultural value issue? Is there still some aspect of the control paradigm and corollaries that do not apply to AL?

As a start, there's a matter of environments. Natural life forms must face controlling the living and non-living environments directly. Belonging to social groups and acceding some of that control for interaction with the environments does not negate the reality that if those groups were not there, the individual would of necessity return to the basic status of directly facing control of environments as an individual life form with no intermediary and completely assuming the task of survival.

The same can't be said of AL forms; they always need some intermediary, human or other AL forms, to obtain needed elemental materials, assemble them, give them AI operational directives, perhaps repair or make new ones, and then disassemble and return them to elemental materials. I can see no possibility for the individual AL form to do this. I could envision some huge autonomous machine performing all these functions, like a massive mobile factory, but it would be impractical. Even a swarm of AL robots working on an individual AL form to assemble and keep it operating isn't the same as what an individual natural life form can do by itself. Our little, natural, DNA-endowed set of molecules performs this function on itself miraculously in comparison as it grows from minuscule zygote to adult human.

It's important to emphasize the significance of the individual control unit. In organic life forms, as mentioned, after the initiation of life, the process of growth and organization of control is done by the individual. In AL, it isn't an individual process but a communal one depending on human input and/or that of other AL forms or machinery. Therefore, the individual isn't the primary and essential focus of control in AL, whereas in organic life forms it is. Certainly, humans, as well as simpler life forms, at times rely on a communal effort to make the individual successful in adaptive environmental control, in the form of social relationships and hosting other life forms such as gut bacteria, but it isn't an absolute prerequisite of existence. Individuals can continue to live without that communal effort and, given enough time and environmental stress, will adapt through structural evolutionary change or, with humans, alternatively through learning and cultural change. AL form individuals can't do this on their own.

This brings up another point of comparison. Since complex neurological processes in the brain have made it possible for humans to have culture (with a reciprocal influence of culture on brain development) as a major form of control, is it possible for AL forms to have culture as well?

We can define culture as learned behavior, concepts, and products transmitted from one individual and generation to another; and this is conceivably possible for AL. The core of AI processing is learning what we could call concepts, organized sets of information, and transmitting that information electronically to other centers of an AL form for execution of directive information—behavior, in other words. That seems to all fit nicely with how it all applies to organic life forms such as humans.

However, learning concepts isn't the same as a judgment in making decisions as to how those concepts are to be used and new ones developed. Computational algorithms for many purposes can be applied to electronic devices where decisions must be made regarding options to be taken. Yet, this is one part of cognition that may be impossible for AI since the capability of summoning off-the-wall jumps to seemingly unrelated ideas for creative, new associations of concepts, let alone assigning socially accepted cultural values to the decision-making, appears beyond AI capability (and perhaps inadvisable to design). Wisdom isn't so easily acquired, even for humans. Still, using AI for the data-crunching and lower-level decision algorithms, which it's so good at, and applying the information to human decision-making would seem to be the best way to enhance the quality of wisdom humans already have.

As to the part about the transmission of culture between individuals and generations, this is easily possible between individual AL forms given their capability (visualizing robotic units here), particularly when it's purely a matter of data transmission. Neither would transmission between generations be outside a legitimate comparative fit with humans, except that the term "generations" with humans is related to time whereas the term with AL is related to a specific level of mechanical and electronic development. Perhaps the term "models" would be more useful—transfer between models.

On the other hand, when more closely examined, the term "generations" as used with humans not only relates to the passage of time but also means that the progeny which emerges from birth is a sort of new model of the human life form by sexual reproduction and evolutionary processes. And, when combined with continuing cultural change (in beliefs, customs, and institutions), the effect of producing a new group of human individuals can be profound.

Applying these same criteria to AL results in the same conclusion: new models appearing over time with cultural influence establishing levels of differences, which can be termed "generational." Even a comparison of the cultural aspect, while somewhat different, still makes an appropriate fit. The cultural influence on individual humans is from outside the individual just as it is with the AL individual. Although the specific type of cultural influence will have differences, it's still an effort to influence changes in the capability and behavior of the individual as a product, whether natural or artificial. Yet, there's one part of that comparison regarding culture which needs further scrutiny: behavior.

Maybe it's just familiarity with the term as used to denote the activity of an organic life form which makes it seem awkward to refer to AL activity as behavior. Also implicit in the use of that term is the nuance that it involves a basic stimulus and response to sensory perception mediated by interpretation as to what should be done, an integration of thinking and acting. Somehow, that just has an uneasiness when used to describe AL.

But shouldn't it include AL? Isn't that what is also the essence of AL: the sensory perception of stimuli via bodily receptors (e.g. light or heat), interpreted by cognitive processing (AI), and directed signals for bodily action (e.g. micromotors)? The difference boils down to the materials used for the life forms: organic for natural forms and inorganic for AL forms. Consequently, it would appear that logically, AL forms can have and use culture, as defined. It would seem that the problem humans have in making such comparisons is emotional, and our sense of self and uniqueness—a problem which AL forms don't have. Now, is this an advantage or a disadvantage, for either life form?

Additionally, emotion itself further needs to be addressed to clarify the extent and relevance of fit between the control paradigm and AL. While some aspects of control are applicable, there's difficulty with emotion. As a complementary and inherent factor along with rationality in human predictive processing, emotion interjects a neurochemical element of change and adaptation into the strictly logical coding of neuronal information processing. This element of chemical effect isn't possible with the inanimate information transmission conduits and body part composition of an AL form: metallic wiring or carbon nanotubes aren't affected as human neurons and body cells are. Granted, in the future, chemical interaction may be possible with carbon, metal, or something similar, but that isn't the issue at stake. It's the effect of emotion that is important, not the delivery system.

In that case, what is it that emotion puts into the human control paradigm? There are several factors which can be mentioned offhand: uncertainty in the initiation and outcome of an action, increased or decreased intensity of behavior, a heightened sense of self-awareness, an integrative function of bringing into action a diverse network of physiological processes and information transfer, and heightened or decreased sensory perception. These and many other such effects are inserted into human control behavior, and all of them are integrated in some manner with diverse cultural concepts and beliefs.

Emotion significantly affects the process and results of an individual's efforts to control environments, whether it's getting mad and cursing the rain that spoiled a long-planned vacation or making sure that a prospective spouse's excellent cook mother appreciates the delicate dessert that was baked for the joint family dinner that night. Emotion is there at some level, in every control decision, for every event of human life. It can be easily said to be one of the most important cornerstones of human identity.

Now, comparing this to AL control, it must be determined what is the same or similar and what isn't. Since the extent of emotion in human control is vast, it's best to look at just a few examples which might provide insight for generalizations. Uncertainty could be inserted into AL control decisions by randomizing AI code processing, except that human emotional uncertainty isn't just random. It's contingent on current and previous circumstances, memory, cultural beliefs, customs, etc., and including the mixture of neurochemicals also thrown into the brew. Could those similarly be factored into AI coding to result in behavior identical or even similar to human emotional uncertainty?

Possibly, to some extent. Would it be worth it, or desirable? It depends on the goal that is sought.

For example, the intensity of behavior could also be varied fairly easily by AI coding, but it would have to be contingent upon uncertainty and other aspects of emotion. Also, a sense of self-consciousness has all sorts of problems to solve if it's to be part of AI emotion since it's difficult enough to understand even for human emotion. Integrated functions of physiological counterparts for AL forms and heightened sensual perception may be more able to tie into the whole of emotion-like behavior, and so forth. But the heart of the matter is still left to resolve: Why bother and what is the goal of the effort?

It seems that in trying to do this, to make AL forms more like humans, the implicit goal is to see if humans can make replacements for themselves. Why would we want to do that? Because we may eventually have so contaminated our planet that no one can survive, and we want something to carry on? Well, if we botched the job so badly in the first place, why would we want another being just like us to continue on and do it again? Or maybe the thought is that we could tweak the replication and eliminate that potential, sort of like what humans have already been doing throughout history with drugs, culture, and possibly genetic manipulation, as previously suggested. In sum, why set out to make a whole new human species based on non-organic material, with an inevitable whole new set of undiscovered problems, when we already have a rather amazing organic model to work with and instead could just try to do a better job with it?

Without a doubt, it's rationally possible that AL forms could help us achieve that goal by humans worldwide working together to determine which elements of human control could best be aided by AI and which should be left to genetic evolution, ecological action, and cultural development. That would be the best scenario and one we could most effectively control if we can apply the optimal cultural concepts and behavior needed to aid in that endeavor. The potentials for both good and bad outcomes from AI and AL are tremendous, just as with all other human cultural concepts; and similarly, it's up to us to achieve a positive balance in outcomes for their use in our future and that of other life forms with which we share this planet. After all, isn't AL just another part of the total non-living environment we seek to control by manipulating inanimate elements into cultural artifacts, be they chipped hand ax, sailboat, or AL robot?

Non-living Natural Elements and Forces

One essential aspect of the control paradigm involves the natural elements and forces that make up the non-living environment on Earth. Earlier, its role was discussed primarily in providing resources for manipulation by living beings, such as soil for termite mounds or stone for human cultural edifices or as locations for habitation such as in oceans, caves, and soil. This environment also provides gases for bodily physiological processes plus weather and climate for maintenance of beneficial, and detrimental, conditions for living. Additionally, there are planetary forces, such as gravity and the geomagnetic field which are the most commonly known and responsible for ocean tidal action and yearly seasons as well as protection from the solar wind and cosmic rays. How the beings making up the living environment interact with these elements and forces is widely varied, but all share a common interconnection with both evolution and ecology in the control paradigm regarding both kinds of environments on Earth.

However, the question being asked at this point in the discourse isn't about living versus non-living environments concerning the control paradigm but is about just the non-living environment and whether or not it fits the control paradigm. Is it appropriate and factual to say that the interaction of the elemental materials and forces of the universe are also seeking control so as to exist?

In general, parlance describing mechanical and electronic functions, it's said that this part of a mechanism or circuit "controls" that part, and so on. It would thus seem that usage of the term could tentatively be extended to planetary systems, star clusters, black holes, galaxies, gravity, electromagnetism, and so on. But does the concept of control as presented here fit for the non-living environment interacting on its own, throughout the universe?

Since the control paradigm is essentially set up to define life, then it clearly would not, since life is a necessary part of the definition. Notwithstanding that logic, perhaps changes could be made in the criteria and interesting conclusions still be reached. Looking back to Part One on Basic Principles, leaving out any reference to a living environment will essentially result in a statement that the non-living environment consists of natural chemical elements and natural forces, that these are all contingent upon each other, and that control of each other to some degree is necessary for each part to continue to exist. This sounds acceptable.

Still, there's a bit of a problem with the term "control." Since the word as applied to living forms inherently includes a condition of the basic individual life form being able to initiate action on its own, this does not fit the idea of an individual unit of the non-living environment, whatever that might be. Conceivably, some individual units if defined right, such as a galaxy or star, could initiate action on their own because of their inherent dynamic forces; other defined units such as elemental particles may not be able to do so.

However, let's not forget that in our discussion of "self" earlier, the point arose that nearly every individual, as initially conceptualized, does harbor a community of other living beings, such as gut bacteria, and raised the question of what an individual life form essentially was.

Hence, a definitional comparison with cosmological objects could be made since a planetary system, for example, has multiple components and that system could be compared to a living system. For the sake of argument, the use of the word "control" as applied to the non-living environment, especially large physical bodies and systems, is acceptable for now.

There's also a clarification needed regarding "parts" of the environment and the term "exist" since the focus of control in living environments is the "individual living being" and "life." Are "parts" and "exist" the exact counterparts here? Not exactly, but they could be construed to be similar. Precisely what the parts are (elemental particles, planet, solar system, comet, star, galaxy, etc.) would have to be defined to be comprehensive. "Exist" would also have to be more clearly defined and seemingly would need to include natural forces, much as a definition of life includes adaptive "forces," or processes, such as evolution and culture.

This brings up the point about life forms being able to adapt and whether a similar meaning can be used with the non-living environment. Certainly, that environment changes constantly and in a broad sense it could be called adaptive; but in the term's use with the living environment, there's the necessity of perception, interpretation, and feedback by the individual so that changes to control can be made.

Consequently, it appears that the analogy with the living environment regarding control would be stretched beyond usefulness, especially when we get to other elements of life forms that involve evolution through genetics, cultural learning, and the transmission of both as key components of adaptation.

It may be best to sum it all up by concluding that there's an interesting broad conceptual similarity but that a direct comparison is impossible because of the basic nature of the units being compared: living and non-living. Yet, the overall idea of control is there, and regarding the non-living environment, humans have been scientifically examining that aspect of Earth and the universe for centuries under organizing concepts such as astronomy, chemistry, and physics. Closer similarities and interactions between the two environments concerning control may still be discovered, with regards to quantum mechanics, for example, and that possibility is worth keeping in mind.

V

SUMMATION AND CONCLUSION

Utility

At this point, we should perhaps ask the ultimate question: What good is all of this control discussion for the understanding of anything, especially in one's daily life?

Some benefits of a general understanding of what makes life function, and how those apply to addressing global and local concerns of which you are aware, have been previously discussed directly or implied and can be summarized as an understanding of:

1. control as the basis for differentiating non-living from living and how that structures the physiology and behavior of all life forms,
2. the human species' place in the realm and interaction of living beings,
3. similarities and differences in how all beings handle control and the requisite resulting conflict that control behavior has,

4. evolution as a key control process,
5. how the human brain and culture are a further development of processes shared with other species and not an exclusiveness,
6. the role of culture in general,
7. specific examination of examples of culture and control,
8. how cognitive processes work, and
9. the present and future role of artificial life and intelligence in human control.

More specifically for the understanding of control in daily life, the following have also been discussed directly or implied:

1. examples from my life and hypothetical parent/child interaction,
2. analyzing the interaction of social organizations with each other and with individuals,
3. analyzing how an individual interacts with their external and internal environments,
4. finding purpose in life,
5. acknowledging the reality that the individual is ultimately in control and appreciating that importance,
6. examining belief systems, informal and formal, for their utility in personal control attempts, and

7. helping decide on courses of action to take by finding and examining the core issues of control in a particular situation by asking:

- What is trying to be controlled, and who is involved?
- What do they want to accomplish; what do I want to accomplish?
- What are the potential outcomes of this control conflict?
- What is the history of interaction by participants, group, and individual, in previous conflicts and similar cases?
- What is the balance between rationality and emotion shown or needed in this control situation, and how is it achieved?
- What is the extent of control shown or needed according to principles and values applied to the situation?

For general control decisions in daily life, besides the considerations above, the idea is to be receptive to cultural concepts and values which seemingly could aid oneself in daily conflict situations. Thus, considering individual freedom, following a middle path, including the total control balance of all living beings, walking in the shoes of an opponent before judgment, and other similar positively valued suggestions are samples of the type of ideas that can be beneficially used in making decisions.

Just on the simple level of common folk values, a short list can be quickly set out as control actions that are utilitarian for individuals. The following is an example of such a list I put together for a young newly married couple seeking guidance in life skills:

> Rules for living successfully.
> Work first, play later.
> If it falls, pick it up.
> Once it's used, put it away.
> Fix it before it breaks.
> Do it now.

Clearly, this will help with controlling household cleaning and management, which was its first purpose, but secondarily it metaphorically applies to individual and social behavior, particularly the middle three: if an attempt fails, pick yourself up and try again; once you're done hashing out an issue, put it out of your mind and carry on; and don't wait to fix a relationship until after it's broken. Even simple aphorisms like these, and hundreds of others available, can have a significant influence in guiding individual control decisions.

However, the purpose of this discourse isn't to construct a systematic set of detailed recommendations but to point out that there are multitudinous examples available on all levels of complexity that can be adopted whole or adapted for one's personal and societal control needs. For example, just a quick search of book, magazine, and website topics of interest and self-help subjects on the Internet provides a wide breadth of information on personal control topics and techniques from which to choose.

Moreover, simply being aware of control and seeing how it's involved in everything we experience can help dispel the uncertainty and misunderstanding of what we perceive in our lives, removing some of the apparent confusion clouding our thinking and emotions. For example, that store clerk appears to snub me, not because of inherent meanness or stupidity, but because she's trying to control the lousy start of her day; those teenagers aren't acting obnoxiously because they don't like me or other adults, but because they're trying to control their difficult growing up processes; that dog isn't growling because it wants to kill my dog as we walk by, but because it's trying to establish a social ranking and verify boundaries of its perceived territory.

A control-infused viewpoint, asking "What are the control issues?" to get to the heart of the matter, can set a lot of apprehensions to rest and allow a more peaceful frame of mind each day. Decide if and how much control you have over a specific situation, take action or not, and get on with your day.

These concluding thoughts aren't a comprehensive summation of what I have attempted to cover in this treatise, nor are they by any means exhaustive in what is possible to include in further study of the paradigm of control and its ramifications for all life forms—especially for humans, since we seem so capable of disarranging the planetary scene in a manner rather disproportionate to our species individual vulnerability. However, they will do for the purpose intended at this time. As stated repeatedly, the reader has all of history, science, and culture to look to for areas of application and elucidation, should she or he be so moved.

Closing

A last reflection and reminder are that although it's intellectually interesting to tease out all the foundational elements of this concept, the simplest test of its veracity is to just think about everything that one does in life, and sees other animals and plants doing, then answer this question: Is there anything you can perceive that isn't directed toward control of something, in either the external environments about you or your own body's internal environment?

What's the purpose of eating that sandwich at noon on your work break and what's the break from? And why did you choose to have an outside table seating with a view of the street activity or the solitude of a park bench where trees rustle in the wind and birds fly overhead? And why are you just now popping that medication pill along with your coffee drink, which you waited in line for five minutes to get?

What's the purpose of that bird flitting about in the air chasing insects? And why is it flying off, beak full of caught insects, to a nest and the chirping sound of featherless hatchlings hidden within it except for the yellow, gaping beaks reaching over the nest edge? And why does that male bird of the same species perch on a nearby branch and frequently fly up and chase other birds which venture nearby? And what's the purpose of its brash screeching vocalization as it pursues the intruder, and why do the nest-bound hatchlings send out a raucous chorus of incessant chirps?

What's the purpose of the tall trees rustling in the wind to grow so tall, while others at the ground level of a different species have larger leaves spread broadly rather than upward? And why do some species of flowers grow in the shady dampness of a nearby small slow-moving stream of water, while others successfully grow out in the sunlit grassy areas in spite of the efforts by human gardeners to keep them mowed down, along with the grass which continues to grow and expand its root system? And why are those flowers attracting insects which birds seek, and also attracting one partner of a human couple who reaches down to pick several of the same flowers and gives them to the companion, as they smile at each other and clasp hands?

I think it's all about control, my friends.

END

ABOUT THE AUTHOR

Lance Packer was born on a farm in eastern Washington state, grew up immersed in a close relationship with that world of structured Nature, and had the time and freedom to think and wonder about what he observed about him during the expansionist times of American life after WWII. He was a youthful sponge absorbing everything newly discovered, from ants underfoot to distant glaciated volcanic peaks to the attraction of the wiles of young girls with flashing eyes. All were approached enthusiastically and sought for further exploration and questioning. Years of further life adventures including Peace Corps service, a Ph.D. in anthropology, and 27 years teaching public school in Alaska broadened that youthful experience. This is the bedrock that has led to this book, seeking to tie all of life together in a simple, understandable idea.

NOTE: Share your thoughts and questions at the author's blog and website: https://www.lancepacker.com.

Also, your honest review posted at your book distributor's website can help spur new readers to investigate these ideas for themselves.

REFERENCES for Further Reading

INTRODUCTION

Maslow, A.H. "A Theory of Human Motivation," originally published in *Psychological Review*, 50 (1943): 370-396. Republished, June 16, 2012.http://www.researchhistory.org/2012/06/16/maslows-hierarchy-of-needs/?print=1

Porth, Eric, Kimberley Neutzling and Jessica Edwards, "Functionalism," *Anthropological Theories*. University of Alabama, Department of Anthropology. Accessed January 2020. https://anthropology.ua.edu/theory/functionalism/#:~:text=Malinowski

I
BASIC PRINCIPLES OF CONTROL PARADIGM

Balter, Michael, "Strongest Evidence of Animal Culture Seen in Monkeys and Whales." AAAS *Science*, April 25, 2013. https://www.sciencemag.org/news/2013/04/strongest-evidence-animal-culture-seen-monkeys-and-whales

Fabiani, Louise. "Animals have Culture Too!" *Pacific Standard*, June 9, 2016. https://psmag.com/news/animals-have-culture-too

Main, Douglas. "Like chess players, these crows can plan several steps ahead." *National Geographic*. February 7, 2019. https://www.nationalgeographic.com/animals/2019/02/new-caledonian-crows-plan-ahead-with-tools/

Ramsey, Grant. "What is Animal Culture." In K. Andrews and J. Beck (eds.). *Routledge Companion to the Philosophy of Animal Minds*. Routledge Press, (2017). http://philsci-archive.pitt.edu/12456/1/Ramsey_Animal_Culture.pdf

II
FURTHER EXPLANATIONS AND EXAMPLES

Barstow, Anne L. "Rape as a Weapon of War." *Encyclopedia Britannica*. Accessed March 2020. https://www.britannica.com/topic/rape-crime/Rape-as-a-weapon-of-war

Broad, K.D., J.P. Curley, and E.B. Keverne. "Mother–infant bonding and the evolution of mammalian social relationships." *Philosophical Transactions of The Royal Society B* 361, no.1476 December 29, 2006: 2199–2214. Published online November 6, 2006. https://www.ncbi.nlm.nih.gov/pmc/articles/PMC1764844/

Cook, Garth with Daniel Chamovitz. "Do Plants Think?" *Scientific American*. June 12, 2012. https://www.scientificamerican.com/article/do-plants-think-daniel-chamovitz/

CNRS. "A single-celled organism capable of learning." *ScienceDaily*, April 27, 2016. https://www.sciencedaily.com/releases/2016/04/160427081533.htm

Davis, Mark A. "Social Behavior." *Biology Reference forum*. Accessed March 2020. http://www.biologyreference.com/Se-T/Social-Behavior.html

Levinson, Stephen C., and Judith Holler. "The Origin of Human Multi-Modal Communication." *Philosophical Transactions of The Royal Society B*, September 19, 2014. https://royalsocietypublishing.org/doi/10.1098/rstb.2013.0302

Lewens, Tim. "Cultural Evolution", *The Stanford Encyclopedia of Philosophy* (Summer 2020 Edition), Edward N. Zalta (ed.). https://plato.stanford.edu/archives/sum2020/entries/evolution-cultural/

Marrocco, Jordan, and Bruce S. McEwen. "Sex in the Brain: Hormones and Sex Differences." *Dialogues Clin Neurosci*, 18 no. 4, December 2020. https://www.ncbi.nlm.nih.gov/pmc/articles/PMC5286723/

Mattison, Siobhan M. "The evolution of female-biased kinship in humans and other mammals." *Philosophical Transactions of The Royal Society B*, July 15, 2019. https://royalsocietypublishing.org/doi/10.1098/rstb.2019.0007

Mulder. Raul, and Michelle L. Hall. "Animal Behaviour: A Song and Dance about Lyrebirds." *Current Biology* 23, no. 12, June 17, 2013: R518-R519. https://www.sciencedirect.com/science/article/pii/S0960982213005721

Ruben, Brent D. "Animal Communication." *Encyclopedia.Com*. Accessed December 2019. https://www.encyclopedia.com/media/encyclopedias-almanacs-transcripts-and-maps/animal-communication

Schmandt-Besserat, Denise. "The Evolution of Writing." Last modified January 25, 2014. https://sites.utexas.edu/dsb/tokens/the-evolution-of-writing/

Scott, J.P., Irven DeVore, and V.C. Wynne-Edwards. "Social Behavior, Animal." *Encyclopedia.Com*. Accessed November 12, 2020. https://www.encyclopedia.com/social-sciences/applied-and-social-sciences-magazines/social-behavior-animal#E

Sukel, Kate. "Sex Hormones and the Brain." *Dana Foundation*, August 2, 2019. https://www.dana.org/article/hormones/

"World's Oldest Writings." *Archaeology*. May/June 2016. https://www.archaeology.org/issues/213-1605/features/4326-cuneiform-the-world-s-oldest-writing

Yonack, Lyn. "Sexual Assault is about Power." *Psychology Today*, November 14, 2017. https://www.psychologytoday.com/us/blog/psychoanalysis-unplugged/201711/sexual-assault-is-about-power

III
DIALECTIC DIALOGUES: Looking Within

Barrett, Lisa F. "The theory of constructed emotion: an active inference account of interoception and categorization." *Social Cognitive and Affective Neuroscience*, 12, no. 1, January 2017: 1-23. https://www.ncbi.nlm.nih.gov/pmc/articles/PMC5390700/

Beckoff, Marc. "Do Animals Know Who They Are?" *Psychology Today*, July 6, 2009. https://www.psychologytoday.com/us/blog/animal-emotions/200907/do-animals-know-who-they-are

Briggs, Saga. "How Predictive Coding is Changing our Perception of the world." *informED*, Open Colleges. September 24, 2018. https://www.opencolleges.edu.au/informed/features/predictive-coding/

Birch, Johnathan, Alexandra K. Schnell, and Nicola S. Clayton. "Dimensions of Animal Consciousness." *Trends in Cognitive Sciences*, 24, no. 10, October 1, 2020: 789-801. https://www.cell.com/trends/cognitive-sciences/fulltext/S1364-6613(20)30192-3

Choi, Charles Q. "Even Non-Amputees Can Feel a Phantom Limb." *LiveScience*, April 12, 2012. https://www.livescience.com/28694-non-amputees-feel-phantom-limb.html

Dom. "The Bayesian Brain: An Introduction to Predictive Processing." Last modified July 28, 2018. https://www.mindcoolness.com/blog/bayesian-brain-predictive-processing/

Geeraert, Nicolas. "How Knowledge about Different Cultures is Shaking the Foundations of Psychology." *The Conversation*, March 9, 2018. https://theconversation.com/how-knowledge-about-different-cultures-is-shaking-the-foundations-of-psychology-92696

Güntürkün, Onur. "The Surprising Power of the Avian Mind." *Scientific American*, January 2020.

Henriques, Gregg. "10 Problems with Consciousness." *Psychology Today*, December 5, 2018. https://www.psychologytoday.com/us/blog/theory-knowledge/201812/10-problems-consciousness

IFLS. "Organisms That Make Us Who We Are." Health and Medicine, *IFL Science*. https://www.iflscience.com/health-and-medicine/organisms-make-us-who-we-are/

Jordán, F., Lauria, M., Scotti, M. et al. "Diversity of key players in the microbial ecosystems of the human body." *Nature*, Sci Rep 5, no. 15920, 2015. https://www.nature.com/articles/srep15920#citeas

Laland, Kevin N., and Luke Rendell. "Cultural Memory." *Current Biology*, 23, no. 17, September 9, 2013: R736-R740. https://www.sciencedirect.com/science/article/pii/S0960982213009329

Loomis, Molly. "20 Things You Didn't Know About...Animal Senses." *Discover*, April 29, 2014. https://www.discovermagazine.com/the-sciences/20-things-you-didnt-know-about-animal-senses

Mescher, Mark C., and Consuelo De Moraes. "Role of Plant Sensory Perception in Plant-Animal Interactions. *Journal of Experimental Botany*, 66, no. 2, November 2014. https://www.researchgate.net/publication/267871302_Role_of_plant_sensory_perception_in_plant-animal_interactions

Michel, Alexandra. "Humans Are Animals, Too: A Whirlwind Tour of Cognitive Biology." *Observer*, Association for Psychological Science, April 28, 2017.

https://www.psychologicalscience.org/observer/humans-are-animals-too-a-whirlwind-tour-of-cognitive-biology

O'Brien, Barbara. "What Are the Four Noble Truths of Buddhism?" *Learn Religions*, April 23, 2019. https://www.learnreligions.com/the-four-noble-truths-450095

Seth, Anil K. "The Neuroscience of Reality." *Scientific American*, September 2019. https://www.scientificamerican.com/article/the-neuroscience-of-reality/

Svoboda, Elizabeth. "Gut Feeling." *Discover*, November 2020. https://www.discovermagazine.com/mind/gut-bacterias-role-in-anxiety-and-depression-its-not-just-in-your-head

University of California - Irvine. "High-resolution brain imaging provides clues about memory loss in older adults: UCI-led study reveals potential tool for early dementia diagnosis." *ScienceDaily*, March 7, 2018. https://www.sciencedaily.com/releases/2018/03/180307132005.htm

Yong, Ed. "What Mirrors Tell Us about Animal Minds." Science, *The Atlantic*, February 13, 2019. https://www.theatlantic.com/science/archive/2017/02/what-do-animals-see-in-the-mirror/516348/

IV
DIALECTIC DIALOGUES: Looking Beyond

Bittel, Jason. "Chimpanzees develop distinct local cultures, and we're destroying them." Animals, *Washington Post*. March 7, 2019. https://www.washingtonpost.com/science/2019/03/07/chimpanzees-develop-distinct-local-cultures-were-destroying-them/

Frank, Adam. "Earth Will Survive. We may Not." *New York Times*, June 112, 2018. https://www.nytimes.com/2018/06/12/opinion/earth-will-survive-we-may-not.html?

Johnson, George. "Wanted: The Meaning of Life." Books, *New York Times*, August 7, 1988. https://www.nytimes.com/1988/08/07/books/wanted-the-meaning-of-life.html

Merino, Nancy. et al. "Living at the Extremes: Extremophiles and the Limits of Life in a Planetary Context." *Frontiers in Microbiology*, April 15, 2019. https://www.frontiersin.org/articles/10.3389/fmicb.2019.00780/full

Rejón, Manuel Ruiz. "Other Forms of Life (Earth)." *OpenMind*. June 29, 2015. https://www.bbvaopenmind.com/en/science/bioscience/other-forms-of-life-earth/

Resnick, Brian. "Slim Mold Plasmodium: How This Brainless Superorganism Thinks." *Vox*, Last modified April 5, 2018. https://www.vox.com/science-and-health/2018/3/6/17072380/slime-mold-intelligence-hampshire-college

San Diego Zoo Global. "Bonobo (Pan paniscus) Fact Sheet: Behavior & Ecology." *SDZG Library*, Accessed February 2020. https://ielc.libguides.com/sdzg/factsheets/bonobo/behavior

Sinapayen, Lana. "Introduction to Artificial Life for People who Like AI." *The Gradient*, 2019. https://thegradient.pub/an-introduction-to-artificial-life-for-people-who-like-ai/

Weintraub, Karen. "Steven Pinker Thinks the Future Is Looking Bright." *New York Times*. November 19, 2018. https://www.nytimes.com/2018/11/19/science/steven-pinker-future-science.html

Wilson, Edward O. *The Origins of Creativity*. New York: Liveright, 2017. https://www.amazon.com/s?k=the+origins+of+creativity&i=stripbooks&crid=3G4DUW625YJ25

References

Wright, Robert. Nonzero: *The Logic of Human Destiny*. New York: Vintage Books, 2001. https://www.amazon.com/Nonzero-Logic-Destiny-Robert-Wright-ebook/dp/B000Q9IRBY

INDEX